A VIRUS IN SOCIETY

THE SUBVERSIVE INFLUENCES DAMAGING HUMAN EXISTANCE AND STANDARDS OF LIFE

Michael J Cole

A Virus in Society

Author: Michael J Cole

Copyright © Michael J Cole (2021)

The right of Michael J Cole to be identified as author of this work has been asserted by him in accordance with section 77 and 78 of the Copyright, Designs and Patents Act 1988.

First Published in 2021

ISBN 978-1-914366-09-3 (Paperback)
     978-1-914366-17-8 (Ebook)

Published by:
    Maple Publishers
    1 Brunel Way,
    Slough,
    SL1 1FQ, UK
    www.maplepublishers.com

Book cover design and Book layout by:
    White Magic Studios
    www.whitemagicstudios.co.uk

A CIP catalogue record for this title is available from the British Library.

All rights reserved. No part of this publication may be reproduced, stored in retrieval system, or transmitted in any form or by any means, electronic, mechanical, photocopying, recording, or otherwise, without the prior permission of the author.

This book is written for my family and friends and for others who have an interest in the subject and those who yet may seek to. Also, to future generations who may be able to judge the value with the benefit of longer term assessment.

The Author

Michael Cole is a successful business investor and holds a Batchelor of Science, Honours Degree in Chemistry and has a passionate interest in the subject of climate and the dynamics affecting changes and the related political, economic and social implications.

This book was initiated by the concerns over the direction of travel of and by societies and the institutional manipulations without mandates to implement change and coercion of the elitist political classes aided and abetted by sections of the media and the uses of modern communication systems. This direction is considered a potential detriment to humanity sacrificed for centralist control. Hopefully it will provide a refreshing perspective on the subjects covered and may be encourage readers to further research the issues and even stimulate action.

A VIRUS IN SOCIETY

Outline Note

This book pulls together other works and writings together with the writers own observations, experiences and research. The views expressed are the writers own opinions.

It covers the considered subversive influences changing our society, structure and way of life in what is considered damaging to our economies in the Western world and so denigrating the social fabric which enshrines our principles of humanity.

These influences have and are becoming more engrained in all areas of our existence and in effect are being forced upon the populous as whole. These moves are designed to restrict freedoms of thought and action and ultimately blunt innovation and development which have been and are the cornerstones of human nature. The rationale behind it is driven by "liberal" views and a key plank is environmentalism, encapsulated by the will to "save the planet". This is a persuasive proposition which superficially is something to be venerated but behind it is what is clearly totalitarianism – this is covered later.

Totalitarianism is the principle of control and domination of humanity and defining the way we live. It is promoted by a group think approach which is now exhibited by many institutions. It is also a key platform of some media organisations and especially the BBC in the UK. A constant drip feed propaganda of how "we" should think and behave reinforces this creeping agenda. People hardly notice the brainwashing!

The seriousness is that no one votes for this, it envelops society in many areas including education, law, politics and the media and even into the very vocabulary so as to become the norm. Those opposing this group think approach become pariahs or heretics as they used to be called in the Middle Ages. Only when the benefits of economic survival begin to fail because of the policies pursued in the name of environmentalism or the concept of totalitarianism will the people see and rebel openly.

The question is can this be contained before it is too late and who and how will it be challenged and reversed from the constraint and demise it is bringing to allow human nature to prosper once again?

This book is one small step to bring this matter to public attention and support others in this direction. Such opinions are open to challenge but they are made constructively and with due consideration.

CONTENTS

Introduction 07

1. Environmentalism (Totalitarianism) 09

    (1) The Basic Theory – Part One 10

    (2) Applied Aspects – Part Two 13

    (3) Implications – Part Three 22

    (4) Next and Final Stages – Part Four 28

    (5) Summary 33

    - General Conclusions 33

    - Main Conclusion 35

    (6) Imagine – 'Green Fascism' – A Story 39

2. Related Assessments – Social Impact Projections 43

3. The Myth of Human Made Climate Change –
   A Selection of Papers 131

4. A Selection of Letters on Relevant Social Impact Issues –
   copy correspondence 159

   Epilogue 273

INTRODUCTION

By personal observation there is a deliberate and effective creeping influence changing the structure of human existence and the way lives are led, especially attacking the established Western cultures and their economies.

This is not the obvious terrorism initiatives and differences caused by religion but an internal, almost psychological 'movement' which can be described as totalitarianism and driven by environmentalism. It is a process of brainwashing. The moves infiltrate every aspect of life using the fear principle or pending catastrophe for the planet Earth and human culpability as the excuse for imposing controls and changes to lifestyle.

This goes further to change and damage the very economies on which human existence relies for its wellbeing and the very nature of the motivation of human activities to develop and prosper. In effect it changes the social fabric for which humans have striven over millennia.

The movement is not an easily identified body, it comprises elements of concepts supported by people who have infiltrated themselves into all areas of structured society including: education, politics, judiciary and the media and who then set and influence behaviour and regulation. They are not elected on any such platform or manifesto, it gains traction by stealth. The public at large don't see it coming but are influenced by the rhetoric and the propaganda largely promoted by the media.

A programme can be seen in the politically correct issues forced upon society, the reduction and control of freedom of expression, compliance with non-challengeable regulations, acceptance of state control (often by the excuse of public safety and the like), non-specific gender issues and, of course, environmental and climate threats. It is the process of group think which takes over evaluation and common sense. The media give much oxygen to these issues with some actually providing a diet of propaganda in support.

This paper attempts to describe the basis of this 'movement' and outline the real threats to human society that it poses. It is clearly ingrained already in parts of Western society under the guise of progressiveness and the question is can it be stopped before it totally dominates the future?

The main section provides a critique of impacts of the concepts of environmentalism and totalitarianism and then a series of papers of comments on social issues related to modern times. The issue of human made climate change has a section on its own given the pending detrimental nature of potential political policies. Included also are examples of correspondence sent by the writer on such subjects to demonstrate a degree of interaction with media coverage.

1. Environmentalism (Totalitarianism)

Of all the social controlling approaches the one set out to appeal to and fixate the public at large is the environmentalist agenda. It embodies survival by claiming to save the planet, it points the finger at Man (humans) for polluting and creates feelings of guilt and fear and presents its motives as a good and noble movement. Who can argue with that?

It does, however, have sinister connotations as when it is examined it is about control, economic reduction and a twisted sense of changing human instinct and behaviour – not for the better.

It is different from other forms of authoritarian control and regimes and it infiltrates all aspects of life in a creeping and unchosen way.

This chapter explores the issues and the dangers and introduces the totalitarian impact of its philosophy.

1.1 The Basic Theory Of Environmentalism

PART ONE

The starting point is the claim to save the planet (from humans and human activities) as a basis for totalitarianism or environmentalism.

This is to promote constraint, restriction and austerity – justified by the, albeit false, theory of planetary existence. The rationale is outlined but it contains more and more sinister moves behind it.

THE TOTALIZING ANALYTIC OF ENVIRONMENTALISM –
The false (but compelling) basis for Environmentalism

If you control carbon, you control life - Robert Lindzen

Based on the theory of Anthropogenic Global Warming (AGW), meaning human caused, environmentalism is now emerging as a scientific theory for which the empire extends scientifically to all things.

There is, in fact, not a single one of human actions or activities that does not generate CO_2, transport, heating, buildings, industry, economy and even the simple act of breathing; CO_2 emissions are consubstantial with the fact of existence, with the very concept of the human.

"If you control carbon, you control life," noted the American physicist Richard Lindzen; upon this truth – a scientific truth – is the empire, totalizing its principle, of contemporary environmentalism.

Authoritarian regimes have marred Western history over time, but totalitarianism is a recent invention.

The citizen of ancient Sparta was accountable to the city for the most binding duties, from childhood to death, Sparta was a military camp, which entailed hierarchy, control and submission of the individual to the imperatives and views of the community. Spartans, however, lived under the rule of law and the separation of powers (Aristotle), thus providing the citizens with a share of freedom (the majority of the people were either slaves or an intermediate category called the Hilotes). The Spartan regime was undoubtedly authoritarian and military, but it lacked the totalizing pretension of the abolition of individuality.

Many "absolute" monarchs desired to rule the individual, but they did not have the means, neither financially nor technologically (and rarely legally) speaking. Above all, the totalizing pretention was simply lacking. The effective empire of Louis XIV over the territory of his kingdom was paltry compared to that of our democracies.

Totalitarianism is a contemporary invention, which emerged in the literature of the nineteenth century, before being implemented in the next century. Totalitarianism is not despotism – the latter is a focal point of figurehead for individuals to kowtow towards – the former is an all embracing concept to subsume individuals.

Fascism was essentially a nationalist form of socialism. Mussolini abolished the elections but left scattered elements of pluralism, including the monarchy. In the twisted workings of Hitlerian pathology, the German was but a cog. The Nationalist Socialist theory endeavoured to conceptually destroy the "bourgeois individualism" of the Anglo-Saxon.

Through re-organising society, communism finally organised everything. Born in blood communism governed arbitrarily. Mao's China, the Khmer Rouge's Cambodia, Lenin's and Stalin's USSR, all these regimes are numbering among the most "anthropophage" of history.

However, communist totalitarianism never planned or even *conceived* of subjecting human activity in its *entirety* to the sanction and control of the State. In the USSR, people travelled as freely as the available means allowed, but travel or transport were never considered problematic *in themselves*. Holidays were taken sparingly, because the means of a planned economy were limited, but if the Party had had the opportunity, it would have increased. Material comfort and consumption were not disqualified in principle, only by the limitation of the capacities of the "red" economy.

Homo Sovieticus was tightly policed, oppressed and materially limited, when he was not being deported in the Gulag and killed. However, he was never considered by the Party and the State as a problem in himself: in his very humanity.

The divergence of environmentalism from previous examples of totalitarianism is not marginal: it marks a fundamental turning point on the essence of being.

If human CO_2 is the problem, then humans are the problem.

This equation is the excuse for more wide reaching issues and implications to support the environmentalists approach – see later.

1.2 Applied Environmentalism -
How it is affecting all aspects of life (and lifestyles)

PART TWO

This section shows how environmentalism rolls out the totalitarianism of its algorithm in every aspect of real life.

This is an anti-human rhetoric to support human constraint and the collapse of democracy and freedom. The programme of infiltration by scurrilous means is used as it cannot be achieved by democratic processes as people will not vote for constraint as abundance and freedom are preferred choices.

1. **Environmentalism and Freedom**

> *The battle for climate is contrary to individual freedoms*
> – Francois-Marie Bréon, Climatologist

> *Coercion? Perhaps, but coercion in a good cause*
> – The Population Bomb, Paul Ehrlich

Thinkers such as Hans Jonas risk advocating a benevolent environmentalist autocracy in the interest of the planet. Environmentalists venture to argue for the abolition of the political freedom that is democracy. No environmental party already advocates the abolition of freedom as such.

Freedom was both invented by the West and is the catalyst and crucible of its development. Conceived by the Greeks, with the concept of *isonomia* (Solon) or equality before the law, taken up and shaped by generations of Roman jurists and publicists; then in

Common Law, by the tradition of the German *Rechtsstaat,* there is the Anglophone Rule of Law, the Francophone état de droit, there is freedom only under the auspices of true law, with definite, fixed and thus avoidable sanctions. Freedom is diametrically opposed to arbitrariness. Economic freedom is intrinsically linked with the market economy and ensures the perpetuation of its various manifestations, including technological progress.

In spite of a century of socialism, our culture remains imbued with the demand for freedom: that value, which is the very condition of morality (Kant).

Demanding the abolition of freedom hardly seems likely to bear fruit and most environmentalists do not. Moreover, environmentalists know that the same objective can be achieved by apparently less radical means. Didn't a thousand small restraints keep Gulliver from moving, as if he were paralyzed from the neck down?

How can we escape environmentalist theoretical mechanics? If human CO_2, is the problem, then human being's many activities are the problem. Will the individual be allowed to go about his or her business as long as CO_2 emissions are inherent to each of them?

Freedom = CO_2

Thus, freedom is being at liberty to emit CO_2 which no coherent environmentalist can tolerate.

The theoretical system by which environmentalists would seize our societies does not allow them to escape this conclusion: in its principle or its applications, individual freedom must be abolished. Let us consider two identical populations. In the first, individuals are free to move, start a family, trade, travel, eat meat, and own pets. In the second, individuals are not free and only allowed activities prescribed by the central authority, for example by issuing CO_2 licenses.

How can we deny that the second group will indeed emit significantly less CO_2 than the first? How can we fail to see the environmentalist theory forces us to qualify the second group as virtuous, whereas the first is harmful, selfish, and "abuse the planet"?

To eliminate ("by 95%") human CO_2 emissions, it is necessary to abolish individual freedom.

In this way, the totalizing analytic of environmentalism induces a totalitarian algorithm:

If human CO_2 is the problem, then humans must be restrained, controlled and conditioned in each of their activities

In an interview with *Der Spiegel* in 1992, Hans Jonas reached the same conclusion: in view of the "ecological catastrophe" and the "technological attacks on Nature," "the renunciation of individual freedom is *inevitable*". This leads to the conclusion that human numbers must be reduced. When and how is the key question!

2. An Ambitious Totalitarianism

Civilization is killing the planet.
Civilization must be destroyed.
Derek Jensen, *Endgame, Vol. 2, Resistance*, 2006

At first glance, environmentalism is a coherent theory. Due to productivism, humans emit ever more CO_2. This CO_2 causes global warming that will ultimately make the planet uninhabitable. So, we have to act before nature collapses because then it will be too late.

One reencounters the theme of the tipping point, constant since Malthus, and the idea of a phenomenon that will take place in such a way that we must intervene in advance because, afterwards will be untimely.

It should be emphasised that contemporary environmentalism does not derive its totalitarian impulse from morality, but from science. When the science, as reported by the Intergovernmental Panel for Climate Change, (IPCC), focuses on the literally destructive and maleficent role of human CO_2, it is natural for those who are concerned about mankind to look at every human activity that emits CO_2, that is the entirety of human actions, from the first to the last exhalation of CO_2.

Real science data or evaluation is put aside by the environmentalists to focus on the elements of rhetoric to support the theory and to reinforce the necessity for challenge as unreasonable and illogical as they may be considered.

In short, environmentalism is a totalitarian Frankenstein's monster that has escaped its creators; if human CO_2 is the problem then humans are the problem. Revitalising this proposition forces us to renounce the theory of Anthropogenic Global Warming (AGW), thus returning to the moral environmentalism that is so weakly built on scientific pretext.

As we see from its stated motives, environmentalism is more demanding, more radical, and ambitious in its desire to subdue humans than any previous doctrine.

Perhaps the closest parallel exists in the *panopticism* of Michel Foucault, on the model of the *panopticon* prison by Jeremy Bentham. Bentham's *panopticon* is a prison structure that accommodates the guard in a central tower around which the prisoners' quarters are arrayed so that they are actually being observed. The control, in action and power, is *total* (permanent). In *Discipline and Punish: The Birth of the Prison* (1975), Foucault generalized panopticism, defined as a desire to impose, through social control, the required conduct on a target group.

One must look to the dystopia of science fiction to see similar representations of total control. The environmentalist pretention, however, is not fictitious.

The totalitarian radicalism of environmentalism can be seen in its identification of "the enemy". Totalitarianism thrives thanks to the enemy. Let us remember the kulaks and other "revisionists" in Stalin's worldview; or the National Socialists regime's perspective on the Jewish community.

Environmentalists do not single out any particular community. Of course they do not approve of the bourgeoisie, the "rich" or the haves but there are no kulaks in environmentalist theory, any more than there are "bourgeoisie" in the sense of Karl Marx. There are only greenhouse gases.

Neither kulaks nor bourgeois; this is because the enemy of the environmentalists is elsewhere. He is in all of us. The enemy is *us* playing our part in CO_2-producing humanity, that is to say the totality of who we are.

From that departure point, a gradation of responsibility (and the degree of hostility) the Westerner produces more CO_2 than the Sub-Saharan African, the White more than the Black, the "rich" more than the poor. He who lives in a detached house will emit more than he who is satisfied with a modest dwelling.

The enemy is in each one of us and no one is deemed innocent; the enemy is humans, in our species, in the very fabric of our relationship with the world.

The enemy is mankind (or humankind).

3. Environmentalism and Democracy: A Broken Marriage - The Creeping Influence by Subversive Means

> Democracy (...) is not the appropriate form of government in the long term.
> Hans Jonas (1992)

Although it is not easy to gain power through promises – there is tough competition – it is virtually impossible to get the majority vote if offering only harassment, constraint and privation.

There is a noted difference between socialism and environmentalism; while the former promises abundance, the latter guarantees insecurity. Elective, noble, moral and even scientific, but insecurity nonetheless.

It is, therefore, hardly surprising that the environmentalist parties remain minority parties and that they have, to our knowledge, never received an absolute majority in any election in any Western country.

Environmentalists occupy 9% of the seats in the European Parliament, a modest and relatively constant percentage of which significant growth cannot be predicted.

Yet, the influence of environmentalists, in law and in fact, far exceeds their parliamentary representation. How is such a phenomenon possible?

It is because, without winning any national election, environmentalists have mastered ideological lobbying and the colonization of places of power at an international level.

The United Nations, the IPCC, the Council of Europe, the European Court of Human Rights (ECHR), the Court of Justice of the European Union (CJEU) are so many small and undemocratic cadres with such real power. International law is binding in the strictest sense. Its influence is somewhat marginal, for instance, in being invoked by judges in justifying a decision. However, by cumulative effect and by self-reference, this quasi-law or proto-law ends up achieving strict legal force.

When a standard is derived from one of these cadres, two scenarios arise. Either the standard is mandatory or applies in each of the national constitutional systems, overriding national law. It is widely recognised that the international standard outweighs the national standard. Put another way, when the international standard is adopted, it is, thus, positioned beyond even the reach of national parliaments.

When the standard is not mandatory as such, it has been noted that it can *become* so, for example by being utilised by lawyers and judges, if only to interpret the actual law. An example is the endless proliferation of UN resolutions and recommendations. This is an apt example because international law is *comprehensive*ly taking precedence over national law. Provided that a modest legal effect is recognised to a quasi-rule of international law, it will wholly prevail over national democratic law; the judges will ensure this.

There are entire sections of law that are being progressively put out of the reach of national democracies and their parliaments.

The case of the IPCC is both composite and of great interest. The third section of the IPCC report is certainly not legally binding; it does not apply with the force of law in our national legal systems, nor does it override, as such, any national regulation. However, this catalogue of "proposed" norms is so detailed and extends well to all spheres of human activity – transport, town planning, construction, economy, tourism, national and international redistribution of wealth – that it is a ready-to-use, off-the-peg product. To be convinced, one only needs to read the third section of the third and fourth IPCC reports; note that many of these "recommended standards" have entered into law.

The case of the European Union is no less singular; here is an international organisation that adopts directives and regulations, strict legal norms that prevail over nation parliaments. Due to its intense regulatory activity – European law is comprised of 160,000 pages – the EU is depriving national parliaments of significant portions of their powers. National democracies are being stripped of their powers that are brought to the institution, the EU, which is fundamentally intergovernmental and undemocratic. This normative alienation is one of the proven causes of the British vote in favour of Brexit.

It is probable at the EU level that the 'green' lobby has become the most institutionalised.

The European institutions are working with many partners and interest groups to help them shape public policy, thus establishing a new model of governance. Among these partners are the 'Green 10', which works closely with the European Commission for the purpose of advice, expertise and assistance in the development of European environmental policies (...). The 'Green 10' is the assembly of the

10 largest environmental NGOs active at European and international level. They assist European policy makers in the development of their policies and are crucial partners in terms of environmental expertise. These organisations receive direct support from the European Commission's Environment DG, which partly funds their actions and supports their projects.

Environmentalists cannot triumph by democratic means any more than they would be able to maintain their position by democratic means; hence, they favour the intergovernmental approach of international law.

In the long term, its totalitarian algorithm condemns environmentalism to consider a form of dictatorship and tyranny, a plan in which a significant number of environmentalist intellectuals are already actively involved. This is the "creeping", subversive approach which is, in fact, being successful – it is not voted upon or even openly declared, it emerges into law and precedent so affecting behaviour and it is difficult to challenge or unwind.

Michael J Cole

1.3 The Implication of the Theory of Environmentalism

PART THREE

Environmentalism is divorced from practical reality or human endeavours with no recognition of the economic needs or of human benefit.

Indeed, it is restrictive in nature of the concept and becomes manipulative in law and control becoming the antithesis of human well-being.

Climate demands outweigh any needs of humans and any leverage will be used to secure this principle.

4. The Non-existent Economy

I've never considered myself an economic expert – Hans Jonas (1992)

From Karl Marx to Lord Keynes, socialism has spawned powerful theoretical systems with real conceptual innovations. For example, the *General Theory of Employment, Interest and Money,* published by Keynes in 1936, masterfully collated the many economic elements into a new conceptual apparatus.

Despite extensive research, it has not been possible to identify a single environmentalist author who specifically proposes a novel economic system, not even a reworking of existing concepts and formats. Environmentalist philosophies, environmentalist morals, political programmes, even an environmentalist macro science, most definitely, but as for a proposed economic system? No.

This is hardly surprising, as economic nonexistence is consistent with the theoretical foundations of environmentalism.

Socialism is an "economism," in that everything is subordinate to the imperative material equality.

Environmentalism is not an economism, because it is defined by its rejection of the economy. From the Meadows Report (1972) to contemporary environmentalism, passing by the so-called altermodalism, it has been as if, tired of their failings in terms of theory, the opponents of the market economy now confine themselves to demanding that it be stopped.

The only economic concept that can reasonably be assigned to the environmentalists is negative economic growth.

Negative economic growth is the idea that we should not increase our economic output, but reduce it; not to encourage trade, but reduce it; not improve productivity, but set it aside.

All ideas that don't make a system. To make a system – that is to say offer a theory, a rational plan – a *General Theory of Negative Growth*, should answer the question of how. Negative growth, certainly, but how? How can economic production be reduced? Which sectors of the economy should be eliminated? Should technological progress, as such, be prohibited, or only its economic applications?

Economic growth means the growth of domestic economic output. As a nation declines, it bears less fruit, a phenomenon observed during major economic crises. The question is: which citizens, individuals and families will see their standard of living reduced? Maybe the elderly, through pensions? Or the "rich" the middle class, or everyone? Is it necessary to introduce a "licence to procreate" to achieve rapid and drastic population reduction?

It is not a matter of arguing that decline is not the goal of environmentalists: it is. However, it is a matter of arguing that such a structure does not constitute an economic system.

5. Greening the Law

Any government decision that could reasonably have an impact on the climate or climate policy must undergo, in advance an impact assessment – Draft of Belgian "Special Climate Law"

Since the climate topic is a question of survival, environmentalists demand that law, value and enterprise are subordinate to it.

This leads them to plead for the adoption of laws and constitutional arrangements expressly aimed at placing climate at the heart of the decision making process, so that no legislative, executive nor judicial decision, is taken that has not been measured in terms of its effects on "the climate" (read: CO_2 emissions).

This is adequately demonstrated by the recent Supreme Court decision to rule the third runway at Heathrow Airport in London, England illegal because it did not factor in environmental or CO_2 emissions calculations in the programme. This, therefore, overturns any economic argument and the democratically voted and approved bill taken by Parliament. The conflict here arises as a result of existing eco legislation arrived at by EU and environmentalists influences as explained in the previous chapter. It is a conflict which will rage on and likely be the focal point for social upheaval.

By constitutionalising "the climate," environmentalists force judges and constitutional courts to balance civil liberties with the climate imperative. This means, in practice, the exercise of freedom is granted, recognised and protected by the law *only if* the CO_2

emissions are deemed "reasonable and proportionate" according to the criteria familiar to the lawyers specialising in public law.

The emergence of the climate into the constitutional arena is an evident paradigmatic shift; it is the logical and predictable juridical continuation of what we have called the totalitarian environmentalist algorithm.

In addition, European environmentalists demand that "climate committees" are set up to advise each level of power in their consideration of climate factors, through regular recommendations and reports. The quasi-normative mechanism that has been so successful at the IPCC is readily recognisable, (albeit that this all embracing, dogmatic approach is questionable on so many levels, including economic, social and "freedom" principles).

It is palpable that this "balancing" of civil liberties and climate demands is just one more step towards a new "civilization."

Knowing that there are no activities or actions of humans that do not emit CO_2 – from the smallest movement to warming our homes – it is foolish to brandish individual freedoms in the face of the collective survival of humanity. Raising the "right" to move, on a whim for frivolous purposes, when the future of mankind is in peril? This is the pretext of the environmentalist's platform.

The tension between freedom and climate is a mere preamble to the ultimate goal of this environmentalist juridical and judicial revolution: the subordination of freedom to climatic demands.

Be it in their moral philosophy or legal technique, we repeatedly encounter the inexorable character of the totalitarian algorithm of ecology: if human CO_2 is the problem, then humans in the infinite variety of their actions must be constrained, restricted and tamed.

6. The Children's "Climate" Crusade

> *"Climate: A Global Strike Orchestrated by Children*
> *Franceinfo,* March 2019

Although convinced of the superiority of their cause, the pre-AGW (anthropogenic global warming) environmentalists generally speaking respected certain limits: they were never seen using young children in their actions.

The new environmentalists do not recognise the legitimacy of these limits. The unfettered exploitation of the sixteen year old Greta Thunberg demonstrates this. The feral use of her to whip up school children's concerns is testimony to the extent to which the alarmists sink.

Now that the survival of humanity is at stake, all means are acceptable and even just; the imperative is both moral and scientific.

Now we see that environmentalists no longer hesitate at having teenagers and even children take to the street "for the climate." In Flanders in January 2019, there was a demonstration "for the climate" of children from nursery school – for 3 to 5 years old – taking up, in chorus, the political slogans of their elders.

Comparisons naturally come to mind. For example, *Komsomol* – those Soviet communist youth organisations – although the Soviets wouldn't allow kindergarten children to enrol; or the Children's Crusade, the lunacy of children fanaticized by adults who ended up being slaughtered by the opposing forces; or consider the enlisting of German youth by the Nationalist Socialist regime, the *Hitlerjugend*. In these three cases, we find the identical stratagem of the ideological subjugation of childhood.

A Virus in Society

In Flanders, kindergarten and primary school children were *obliged* to participate in these events (as part of their compulsory schooling), even if this was contrary to the opinion of their own parents. This politicization of childhood, contrary to the family's wishes if need be is not common in our democracies.

Can we imagine schoolteachers forcing four year old school children to demonstrate for capitalism, socialism or Christianity? Why does what is absolutely unthinkable become legitimate when it comes to environmentalist ideology? The legitimacy of contemporary environmentalism rests on moral considerations, like an ideology and doctrine, but that this ideology mesmerizes with the alleged endorsement of science radically distinguishes it from its rivals.

This comparative advantage allows it to nurture global ambitions and demand that the obstacles – freedom, economy, pluralism and childhood – which stands in its way be dismantled without delay. The plans here are deliberate and organised.

Does the survival of humanity not deserve certain sacrifices?

Let us now turn to the phobias of environmentalist ideology.

1.4 The Next and Final Stages of Environmentalism

PART FOUR

The result of the theory, after putting in place a full restructure regime and legal structures is to end economic progress of any kind and force humanity to adopt and adhere to a different existence.

This will be one of global governance. It should be careful, however, as humans will likely revolt eventually – but until sanity dawns the period between will be catastrophic – unless common sense and recognition of the real science and human needs prevail before the environmentalists get their way.

The future?

Environmental Phobias

> *For the climate, please avoid the car, meat, and children*
> Lorrie Goldstein

> *Ideally, we should abandon the plane. Give up your personal car, do nothing but walk, cycle, take public transport and, therefore live in urban centres.*
> Mélanie Gelkens

Let us see where the totalitarian algorithm of their doctrine leads environmentalists in addressing the detail of the entirety of human activity.

Elimination is the principle of environmentalism.

It is to eliminate (and reduce the numbers, influences and very nature of) humans and to focus on all aspects of human behaviour (and quality of life) which has been developed and to which the world at large aspires.

These attacks are on:

- Travel — of all types — planes, cars, rail, goods transportation — blaming CO_2 and a will to eliminate human made contribution. But the real reason is to eliminate economies.

  No real alternative has been proposed as swapping fossil fuels for electric power generates the same or more CO_2 levels (not that CO_2 is a bad thing, it instigates propagation in plants and food production AND has no significant impact on the planet's temperatures as can be demonstrated by historical real world scientific data).

  The proposal to have any travel through 'public' or governmental control systems is the environmentalist solution. It will be inefficient and not CO_2 eliminating — it is totally impractical to do this.

  Such moves would ruin economics as business and trade would cease or at least private enterprise would. This, however, fits the model and motives of the environmentalist movement.

- Meat — elimination of meat consumption would be affected — again blaming CO_2 as the rationale.

  This is more to do with the environmentalists' mantra than practical matters and it is another example of totalitarianism.

- Agriculture — a further elimination of modern agriculture is demanded by the environmentalists' philosophy.

As specified toward meat consumption and the elimination of animal rearing for food, so is the approach to modern methods of land management. A return to rudimentary methods is the proposition on the basis of less CO_2 production.

The problem is that productivity and food supply will be reduced and the question raised is who will go hungry? The banning of pesticides will further reduce yields and the hiatus of supply will bite especially in vulnerable areas around the world.

No doubt this will help the move to reducing the human population as part of the environmentalists aim!

- Nuclear Energy – a non-CO_2 form of energy so you would think environmentalists would welcome it – but no. The ideology of theirs falls down.

There is a fundamental drive in the philosophy to eliminate any form of industrial process and economic success. The deep motive is to constrain human endeavour and ultimately reduce human kind. What else explains their position on nuclear power?

So-called 'renewable' or 'sustainable' power and energy sources are, at present, technologies (and they wish to restrict technological advances so the future looks limited) and are not capable of supplying needs without fossil and nuclear generation so again eliminating current methods will leave shortages. Life as humans know it will be dramatically changed and so damaged.

The alternatives of wind power, solar energy through solar panels and the like require huge amounts of energy to produce with CO_2 emissions as well as being unreliable by nature.

The solution would be less humans!

- Population – this is considered the central cause of CO_2 generation and, therefore, humans are the central problem – the number of people and their life support requirements.

  The core issue, as conceived by the Club of Rome and Paul Ehrlich's book, "The Population Bomb", clearly states the need to control and even eliminate people to reduce numbers.

  This cornerstone philosophy sheds a light on the real motives of environmentalism. A dangerous and dramatic concept which if undertaken will deliver George Orwell's worst nightmare.

- Summary

  In summary, the environmentalists' principles and strategies for change are driven by a controlling and restrictive approach to human existence. The excuse used is the environment and "saving the planet" from a human made catastrophe caused by CO_2. They blame CO_2 and relate this to human existence and, therefore humans are guilty.

  Despite the fact that the science does not support the claims it is the justification used to propose the changes to structures of human existence, economic development (or elimination of it) and social rules under which humans should be managed.

  This is akin to the Marxism approach of "Critical Theory" and whilst this raises another discussion it is a close mantra to infiltrating society, particularly in the Western economies, through influence of education, political and judicial institutions and, of course, by media manipulation. Critical theory is explored later.

  The propaganda is well advanced and has infiltrated influential organisations and the media. An awakening is needed before this philosophy gets hold like a religion and those challenging it become heretics. The question is; can this be stopped and how?

A programme to make the public aware of the real science and an influential body to promote the true facts and explain the dangers is needed – and soon!

1.5 The Summary Tenets of Environmentalism

GENERAL CONCLUSION

> *It is the people who enslaved themselves, who cut their throats.*
> Etienne de La Boétie, *Discourse of Voluntary Servitude (1959)*

If human CO_2 is the problem, then humans are the problem

Whichever path we choose, theoretical or practical measures, contemporary, environmentalism brings us back to this truism, this obvious truth. If human CO_2 is the problem then humans are the problem.

It is worthless to object that the "polluting" industries are the issue at hand and not individual daily activities.

Transport is one of the main causes of human CO_2 emissions. Such statistics are only the sum of a multitude of individual choices. Container carriers that sail from one continent to another are the materialisation of individual free choices and the imposing incarnation of human activity at large. The same applies to heating, construction and other human sources of CO_2. Travel and air travel is the culmination of freedom and used to promote commercial and economic enterprises for the benefit of human existence and to constrain or eliminate such activities would clearly be detrimental to human existence.

If given free rein, human activity generates CO_2. It is intellectually impossible and practically inconceivable to isolate those human activities that emit CO_2, those that could be curbed and supressed, and those deserving of carte blanche.

To compensate for past CO_2 emissions, humans must not only moderate current emissions, they must reduce them to nothing; the IPCC hath spoken. Certain political parties, in the afterglow of that IPCC bombshell, no longer hesitate at advocating the reduction of CO_2 emissions by 95% within 20 or 30 years.

This objective can only be achieved by utterly dispensing with human freedom; because:

Freedom = CO_2

This is what is called the totalitarian environmentalist algorithm.

If human CO_2 is the problem, then humans must be controlled, restrained, and brought to heel in each of their gestures and activities.

The totalitarian algorithm shows itself in the environmentalist academic literature, as the demands of environmentalist organisations, and as the practical proposals of environmentalist parties. Totalitarian proposals, which suggest, recommend, and demand that humans be restrained in each and every activity; travel by car, travel by plane, meat consumption, energy, pet ownership, euthanasia, (un)caring for elderly, and even not giving birth and forced sterilisation.

In effect the CO_2 attack is a hook, a con, an excuse to pressure environmentalists' real aims or philosophy of reducing humans and eliminating human activities, economy and lifestyle and to subrogate humans to the Planet (or Nature).

Environmentalism – Summary Review

MAIN CONCLUSIONS

To limit the spheres where freedom of choice can be exercised is to undermine the very essence of man.

Isaiah Berlin

1. Environmentalism is not a sub-branch of Marxism – it preaches totalitarianism and is ambivalent to humans, other than limiting their effect on the Planet.

2. Malthusian predictions have been made on the theme of resource depletion for the past 250 years – Malthus, Jevons, Ehrlich, Meadows and the Club of Rome participants etc – have all been proven wrong, without exception – and including the IPCC modelling of climate predictions.

3. Environmentalist ethic is "physicist" and anti-human species: environmentalists wholeheartedly favour nature without discrimination and humans are subordinate to the 'needs' of the Planet (or Nature).

4. If human CO_2 is the problem then humans are the problem (totalising environmentalist analytic) – despite real world science data the proportion of CO_2 as the 'bête noire' has been singled out by environmentalists.

5. If human CO_2 is the problem, then humans must be restricted, constrained, and governed in all their activities (totalitarian environmentalist algorithm).

6. The abolition of individual freedom does not "derive" from the totalitarian environmentalist algorithm: it *is the* algorithm –

therefore freedom and activities relating to the likes of travel, heating, wealth and even procreating and ambition must be sacrificed.

7. The homicidal tendency of environmentalism is consistent with its philosophy.

8. The de-population of the Earth is the final solution to the environmental problem.

9. As a science, ecology should not seek to enter the political arena; when it ventures into it, it ceases to be a science; the climate change 'movement' and environmentalism through the IPCC and UN in general has done just this to the detriment of 'real' objective science.

10. Environmentalism is a form of totalitarianism.

Each of these propositions is structurally refutable. Propositions 3 to 8 have been fully adopted by a significant number of environmentalist intellectuals.

Key Points as Summary

1. Environmentalism has false premises to promote its theory to "save the Planet" – scientifically unsupported basis to blame humans for the proposed destruction of the Planet – equally wrong basis to equate CO_2 as a destructive medium.

2. Environmentalism's philosophy means totalitarianism to control humanity, its numbers and activities on the false premises which effectively would destroy humanity and its wellbeing by abolishing ambition and development which are inherent in human nature and they would even wish to promote forced extinction programmes.

3. The promotion of environmentalism is carried out by a subversive and "creeping" approach by infiltrating institutions and altering laws and behaviour to shape its philosophy in the public's mind without any vote or open consensus.

4. The result of such changes to social structuring will destroy economies and balanced societies by putting human existence totally subordinate to the claim of the Planet's survival and human influence upon it (even though this claim is only rhetoric and unproven). The relevant decisions in this new world order would be made by the ruling elite in a world devoid of democracy in any form.

Comment

As inferred and expressed throughout this paper the environmentalist's agenda is focused solely on what the supporters believe or at least portray as the "needs" of the Planet for its survival. It regards humans and human activities as the reasons for such a threat and wish to demand that humans are controlled and their activities curtailed, in a way which requires the ultimate of depopulation.

This is totalitarianism, an all embracing governance of human behaviour and activities to make them subordinate to their defined needs of the Planet.

This is a philosophy akin to a religion effectively divorced from real, objective science or scientific data. In fact the real data defies the premise or need of such actions or the basis of such a philosophy.

The cornerstone of the environmentalist's proposition is that CO_2 production from human activities and presence is the cause of the damage to the Planet. The data does not support this and the philosophy and rationale for environmentalism is, therefore, wider and arguably more sinister than is admitted.

The philosophical approach and subversive introduction into the structured lines and institutions of human organisation has already begun and if it continues it poses a real threat to the ways of life, quality and satisfaction of existence and even the natural desires of humanity. Ambition and development will be sacrificed for the vague and unnecessary needs of the Planet as they claim and for the totalitarian principles of this philosophy. A world governance by an elite group is the prospect and one which could be considered unimaginable.

It is imperative that this is seen for what it is before it creeps into every aspect of human activity and destroys the values which are part of existence. It is difficult to engage or tackle an "invisible" invader apart from resisting each move when it is evident even if this means appearing heretical against the group think position which has infiltrated organisations and bodies who support this philosophy.

The real challenge needs to be faced now in the year 2020 before it gains more footholds and this is, not putting too fine a point on it, for human survival! It is a potential self-imposed destruction.

1.6 IMAGINE -
THE RESULT OF ENVIRONMENTALISM
– a human story
"GREEN FASCISM"
coming your way soon!

Michael J Cole

A FICTIONAL TALE OR
A PREDICTION OF AN
ENVIRONMENTALISM FUTURE

*A story of a boy caught in the world of "Green Fascism"
– a post environmentalist apocalypse*
Matthew Meditates

In the confinement of his cell, Matthew meditates. There's not much else to do, except the physical exercise, which he forces himself to perform. Matthew just turned 13, but no one thought to wish him a happy birthday.

Matthew wonders how previous generations saw the world. How a child – a young teenager – of his age might have seen the world, the past, his future. Matthew can only see the aftermath of the upheaval.

It all began in a minor way, and under the benevolent eye of science. It started drastically reducing car traffic before banning it by collectivising transport. The crime against humanity of denying Man's depredations of Nature was made law. Article Zero, the Climate Article, was given precedence in the European Convention on Human Rights, the imperative of the survival of humanity overshadowing civil liberty. The new article was adopted in most national constitutions. Environmentalist economic planning offices were soon set up to ensure that no economic initiative or activity took place that contributed in any way to global warming. In France and Belgium, Article 544 of the Civil Code, which enshrines private property, was not repealed as such but was somehow supplemented. "Property is the right to enjoy and dispose of things in the most *absolute manner, provided that they are not used in a way prohibited by the laws or statutes"* became *"Property is the right to enjoy and dispose of things within the limits of the law and climate imperative."* High-speed internet – what was then called 5G – was outlawed, in the name of public health and the preservation of resources.

The combined effects of these measures was called Environmental Deceleration, and, soon, the Age of Modesty. Unfortunately, this modesty did not last, because soon after, the economy collapsed and the unrest grew. Oliver Cromwell's regime looks joyous compared to this!

Increasingly violent disorder justified the establishment of a state of emergency. Pensions went unpaid, social security benefits were scrapped, the courts ceased to function, and travel and heating became difficult since everything had been electrified and the national energy grid had collapsed under the weight of its own costs. The elections were postponed, then suspended, and never reinstated.

Faced with the threat of chaos and anarchy, the ineffective national administrations were gathered together in a European Sustainable Economy Plan, with national "chapters".

Matthew does not know the details; when his father speaks of that time, he does so with passion and even fervour. All Matthew knows is that due to the European Sustainable Economy Plan and the establishment of the Nomenklatura of Resources, the population of his village shrunk by 80% and he is now the only child.

How did we get here? How can we justify that the Nomenklaturists live in luxury – luxury that Matthew had thought only possible in the bygone era – while he and his fellow men are poor. Yes poor! Miserable even! Everything rationed: water, electricity, heating, CO_2... Where am I now?

Michael J Cole

"Why?" thinks Matthew. Should I solve it? Am I nothing? Just another resource? Is there nothing in me that justifies a form of ...? The child cannot find his words.

He awaits the verdict.

ಜ‌ಜ‌ಜ

2 RELATED ASSESSMENTS -

SOCIAL IMPACT PROJECTIONS

A series of papers by the author on social commentary are included in this section outlining issues having an impact on social constructions and human interactions – set in the 21st century Western environments.

These constructs are having a material effect on human behaviour and the way society is directed, encouraged and managed. They are being incorporated into law and education fostered by politics and a never ending force of political correctness. It is promoted by large sectors of the media and especially influential organisations like the BBC in the UK as they use propaganda to reinforce their narrative and view of how they wish things should be.

These approaches are not necessarily what the public at large want but the drip feed gradually resets the standards.

This section points out some of these and the possible dangers they engender.

NOTES:

The term "woke" is defined by the writer as: 'focus on attitudes towards social issues expressing excessive sensitivity and demanding compliance by all to such views held at the exclusion of alternative opinions'. It is restrictive in its approach or tolerance to debate.

The "PC" term, political correctness, is applied by the writer as: 'to express views on subjects to conform to a pre-determined mantra designed to avoid upset, confrontation or challenge and to fit with an 'accepted' position'. It does not support tolerance of alternative views or debate and demands compliance.

Both these terms have a derogatory effect on free speech or real liberalism and they do not acknowledge that varying views should exist as conformity is key. This approach is considered restrictive, illiberal and an anathema to the proposition of free thought. It is constraining by action and centralist by control.

We are all being guilt-tripped by the woke and PC brigade – guilt for climate change, guilt for our ancestors even guilt for being human and for our species ability to develop.

Woke engenders guilt. We have come so far from living in caves and some will have us throw it away!

TOTALITARIANISM
– the challenge to existence and real freedom

What is Totalitarianism? It is the drive to purge any form of democracy and to not only put controls on society but to constrict any activity of humankind in conformity of principles of a selected mantra.

This mantra is Environmentalism. This is formed on the basis of any form of human activity that is considered by those promoting the concept to be a contributing factor in the destruction of the planet Earth.

This puts humankind as the guilty party. By undertaking any human activity such as transport, heating, building, industry, eating and even breathing it creates carbon dioxide, CO_2, and, therefore, Environmentalists claim it is destroying the planet. The very existence of humans is, therefore, to be controlled or even eliminated.

This may sound fanciful but consider the existence of the Club of Rome, formed in the 1960's with the stated principles of controlling population, sterilisation programmes and even forms of euthanasia. Paul Ehrlich's book, 'The Population Bomb', expounds the dangers of population growth and resource depletion to support such concepts.

Although the principles behind these concepts were wrong as were the predictions of social and human collapse, there are elements of this thinking in the Environmentalists approach to the much distorted theory of Human Made Climate Change (previously known as Man-Made Global Warming).

These tenets of blaming humans for the potential destruction of the planet are used to justify the move to eradicating human activities, to suppress CO_2 production and "save the planet".

Apart from the incorrect interpretation of the science, ignoring established data and the misunderstanding of the CO_2 effect and its natural balance it is the deliberate manipulation of the data and distorted presentation through the media to the public that is the catastrophe. Promoted by those with vested interests and supporting the concept of Totalitarianism such as the IPCC and members of the United Nations, the "education" of humanity has been conducted. It is or is close to being a religion which cannot be challenged and is reinforced by propaganda. There is a constant diet fed to the public and the BBC is a particular protagonist of this (which now prohibits any challenge to the concept of human made climate change).

It is becoming the equivalent of the Middle Ages edicts which for those challenging them were considered heretics.

Totalitarianism is the most serious challenge to human existence and freedom.

Totalitarianism is different from Authoritarianism in that the latter, whilst exercising control over people and being despotic in nature, it is largely 'local' and it has in past examples such as Communism and Nazism accepted the existence of people (even if it killed many), Totalitarianism is a global concept which challenges the very existence of humans and their activities of existence. It also has technology to further exercise control and hits at the very nature of humanity. Enterprise, development, freedom, choice and even ambition is to be sacrificed at the altar of Environmentalism – the planet overshadows any need for humans and so humans, society and wellbeing must be subrogated to the higher principle.

This will result in real economic decline, social collapse and a backward world for human existence in the extreme. It will require a global authority to oversee and manage the human structures like some extrapolation of George Orwell's books.

If the so-called platform of human made climate change and Environmentalists theory are allowed to be secured and Governments let it change our fundamental principles of freedom and development then it will be catastrophic. Economies will fail and social enrichment falter.

A fundamental change and resistance to the false god of Environmentalism and its effects of limiting human endeavour needs to be effected – now!

The effect of human activities on the galactic arena is infinitesimal and to use the inaccurate presentation of data to empower the human guilt theory and change our character is misplaced in the extreme.

Ref: Club of Rome/Paul Ehrlich – see publications

Michael J Cole

ENVIRONMENTALISTS AGENDA – "FAKE NEWS"
A Programme for Disaster

Why have so many people and in particular, politicians, adopted the campaign of 'Saving the World' buying the 'Climate Change', previously 'Global Warming' (until it did not transpire), message?

True, it's because they by and large have not researched it, have been brain washed into accepting the 'religion' and want to belong to the crowd as well as seeing a financial benefit for tax enforcement.

But the politicians see this as a distraction to tackling local or their country issues by 'facing up to a world catastrophe' in a Hollywood type film, leading people to safety. Frightened people need leaders after all and such a large galactic challenge effectively stops them being judged on more mundane, but arguably more pertinent issues on which they can be challenged. Global issues are more diverse and require intergovernmental cooperation - a good excuse when they fail.

The real concern and disappointment is the lack of real understanding of the science. The distorted presentation of the data to support the climate change lobby's predictions is allowed to go as 'settled science'. It is not and the true data gives a different picture. This is serious as the politicians who direct policies are sleep walking into economic ruin.

How has this situation occurred? How have the institutions and influential bodies been persuaded to follow this distorted 'evidence' in the face of real world data and created a dogma to direct policies?

So called scientists, (so called because they lack the essence of science which is objectivity and evidence based evaluation) initiated this concept for notoriety and then seeing the opportunity, money! Vast sums of research grants in trillions of dollars have been doled out on any limp proposals relating to climate change research.

Then, as with many examples, the crowd following principle has applied. Public fear has been created and alarm bells kick in and people want leaders to solve the problem. A problem which doesn't exist by any reasonable evaluation of historical data, CO_2 balance and the real world measurements.

All the science is lost in the rhetoric and head-line sound bites. So called celebrity support adds to the momentum and endorsements of the propositions – such celebrities are largely ignorant of the true science and want media coverage for 'saving the planet'.

Coupled with this is the deliberate act of vetoing any challenge to the 'settled science' by the media and organisations supporting the 'religion'. The BBC is a classic example and is constantly promoting the propaganda of the man-made climate change creating guilt in its viewers and listeners. Deniers of the misguided theories are heretics and not to be listened to. Indeed, the BBC agreed a journalistic edict that ... the consensus was so compelling no equal air time for opposing views should be allowed! This, from a public body, whose charter was for balance and impartiality. And so they have done this by suppressing any challenge to the dogma of man-made climate change and the need for urgent action to change our lifestyle.

So, the real science data has to be reported in a deliberately distorted way, the promoters of it have personal and financial reasons for doing so and the political establishment blindly go along with it. Additionally, the public are denied the opportunity to consider opposing views by the media blackout.

There are many publications challenging the concept of man-made climate change which convincingly demonstrates the frailties of the concept and the misconstrued scientific data and unfounded projections. Most people won't read them – they should. The politicians and policy makers undoubtedly should to ensure they understand the real data and avoid making the wrong decisions.

This leaves us, the public, and mankind, with the prospect of economic collapse on the basis of an "Emperor's new clothes" theory. Time will, of course, tell if the theory was false, by which time most of us living today will not be here.

It is likely history will judge this concept of our time as a misguided period and the biggest confidence trick of all time. But for those alive today and in the near future who suffer from the policies imposed by this it will not appear funny.

Let's now look at some basic information to lay bare the man-made climate change theory.

There is no real problem of CO_2, the value and concentration levels are well within manageable proportions and amounts to 0.04% of atmospheric gases (excluding water vapour) and water vapour, uncontrollable by man, has the largest impact on the greenhouse effect. CO_2 actually promotes vegetation growth and at 400ppm currently is less than the average over 600 million years. The threshold for vegetation survival is 150ppm below which vegetation cannot survive.

Also, the man-made claim is not borne out by the real data as a significant warm period of the Earth in the middle ages around 950AD to 1250AD is well known when no industry existed and no cars were used!

Then there was a cooling period around 1450 to 1750 when temperatures dropped.

During the phase after the two world wars CO_2 production increased with industrialisation development but the global temperatures continued to decline until around 1977 so the correlation of CO_2 and the Earth temperatures is clearly not proportionate. This further demonstrated there is no linear connection between CO_2 and Earth temperature.

The warmist supporters would have us believe that CO_2 is currently so high and damaging that the Earth is in imminent danger of burning up – it isn't!

Temperatures also then stabilised and remained flat from 2000 to 2017 when again CO_2 production increased.

All this shows is that there is no evidence of CO_2 causing warming. It also demonstrates that other factors are now in play relating to the Earth's temperatures – and any related climate effect. The solar effect has most impact here, unrelated to man-made issues.

In summary the connection with man-made impact is not proven, in fact it is disproven.

There are inconvenient facts which challenge the basis of the climate change argument and precisely the man-made effect. There are many other factors undermining the theory.

The so-called climate change scientists and including IPCC and supporters have taken selected parts of the historical and established data to reinforce their theory and often, as proven in the case of East Anglia University, deliberately manipulated the data to fit their arguments. The computer modelling was also constructed to provide projections to suit. This was exposed in the case of EAU and the IPCC presentation which they had to admit in the infamous 'hockey stick' presentation deception. However, this distortion has continued and the rhetoric has taken over from fact.

So, we have clear evidence to defeat the man-made link theory and evidence to discredit the presented data as selective and distorted which has been used to support the theory.

There really is a need to revalue the whole philosophy of climate change and stop the rush to destroy our economies and way of life in the developed and developing world.

Quoted data based on publications from:
US National Weather Service
Barnola 2003
Berner 2001
IPCC 1990 (modified)
And referred to in publications by:
Gregory Wrightstone
Christopher Booker
James Delingpole
Bjorn Lomborg

HUMAN GUILT! - CONSIDER

To hear the environmentalists talk and the social commentators you would think humans had no right to be here, on the planet Earth!

Yes, be respectful to other forms but we have no need to apologise for our existence.

The way of behaviour is important and attitude towards our environment and to each other human. Some of it will be harmonious some will not. Some will live within human laws some will not and the codes of living – laws and ethical standards – will differ across many areas, nations and communities. But we cannot and should not feel guilty about our existence.

The, what appears, constant criticism of human impact on the planet is unjustified and waiving our right to exist and live and improve our standards of life. Enjoyment and survival is an inherent gene within most of us.

The planet survived well before humans (in our present form) and will do so long after we have gone. If we go it will be without a trace over time. All we have produced and created will diminish. For the vast majority of us no one will know we have been here in 100 years from now!

The planetary system and cosmos is far more resilient and forceful than mere humans – we are but a short lived and temporary occupant of Earth and space. Let us not be arrogant enough to think beyond our position – as a very, very small speck in a vast universe.

So, stop this self-critical defamation of humans and human nature!

Michael J Cole

THE FUTURE WE ARE FACING
..... being misguided in the present

Influences in our lives in the developed Western world are self-generated and are becoming restrictive and damaging to many of the principles we, especially in Europe and particularly the UK, held as a foundation of our social structure.

These principles were reinforced after the First and Second World Wars. They included freedom of speech, freedom of thought and expression, open mindedness, fairness and humane and reasonable treatment of people. Extended elements of these principles covered tolerance and equality, including gender preferences, race and backgrounds of social upbringing.

Much of these principles were enshrined in constitutions, laws and operating contracts which became logical developments of reasonable behaviour of a developed society.

However, the present position and accelerating change of some of a more narrow set of interests have begun to distort the reasonable and arguably sensible platform causing disruption and challenges to the accepted standards so upsetting the balance. Such imposition of new and unbalanced concepts are used to beat submission and attack non-believers.

Platforms in the media and, of course, the internet give opportunities and feed oxygen to the fire for promoting and advancing such concepts. This is done to the point of absolute 'truth' and a dogma which for those who don't accept it they become heretics – to isolate and force surrender through bombardment.

What is lost here is objectivity and the acceptance of alternative views – even in the face of opposing evidence. The religion of the 'new' idea becomes all-embracing and dominant.

Areas of this type of indoctrination includes climate change, gender inclination (even assigned to children as young as pre-teens), social origination and politics, drugs and even Brexit with the divides it creates.

Such challenges eat into the whole basis of the populations' life style, economics and the notion of democracy.

The downside of all this is the deleterious effect it has on social structure and interaction of people, individuals and groups.

Having no ability to challenge these concepts and no acceptable reason to do so by the people supporting them causes conflict. It's not just the idea or concepts but the process that now such ideas and concepts become part of everyone's lives whether they want to agree or not.

Often, it is a minority view to start with, picked up by the media, especially the BBC, who then repeat and expand it under a 'liberal' banner and soon such concepts become the 'background radiation' and take on respectability by the liberal elite. Such liberal groups are anything but as they don't accept views not consistent with their own.

Where does this lead us?

At one time the change in world economic balance was thought to be a potential driver for social unrest. It still is a likely cause that the 'West' loses out to development from the Far East, India and Africa, once it begins to take hold. When established economic and social structures are challenged and if they begin to diminish relatively to advancing economies and people in the countries do not enjoy the economic or social living standards that they have become used to then it sows the seeds of dissatisfaction.

If these societies hit the trigger when interruptions to essentials of modern life occur such as food, energy and other basic substance of Western life then the trouble will start.

There is a trend that such a scenario is beginning, even though the West, USA, Europe (and the UK) are still operable as economic dynamos, future trends may not be so rosy.

And, this is a mounting potential problem with the changing social restrictions mentioned above and the limiting democratic influence felt by society as a whole through politicians ignoring people's wishes and the non-allowable expressions of views enforced by law (eg hate crime legislation – when none was intended) and the conformity required by the dogma approach of this new 'culture' and this will force people to react.

Couple this with economic and living standards decline and you have a powder keg waiting for ignition.

Most revolutions, in historical analysis, are started by intellectuals and ideas. The numbers are provided by the aggrieved masses who feel they have been wronged. In the Western societies we have both intelligent people and large numbers in middle class structures – a combination of great potential power and influence if it can be combined. All it needs is a cause and a leader. If the new status becomes so poor in terms of economic/lifestyle standing and such restrictive controls with little or no choice then people have nothing to lose so action is the solution.

A Virus in Society

Evolution is change over a long or controlled period and can be managed and accepted. Revolution is rapid change over a short period and usually occurs when particular circumstance coincide in a dramatic fashion where a common cause is clear. The signing of the Magna Carta in 1215 was an example. Many others exist.

In today's modern world mobilisation of forces happen differently but the basic and fundamental causes are the same; mass public dissatisfaction, a common cause and the governing elite at odds with its people.

Predictable dates for such conflicts, internal self-generated destructive issues and exacerbation by outside influences, suggest within 100 years if we do not resolve the social restrictive agendas outlined here and if economic decline strikes if we in Europe and separately in the UK don't successfully compete in World trade terms then within 50 years could be likely.

Because of these artificial and false culture restrictions rapidly forced upon our society and the equally draconian restrictions being applied to industry our economic performance will be impaired to make us uncompetitive and with the consequences this will cause adverse economic impact and damage to our social and life value standards.

Restrictive freedom, economic decline is a recipe for disaster. The public will fight back. How it does this will depend on the governing bodies' reaction.

Violent it could be, aggressive in the rate of change it will be. Those in governing power and the media must realise the status of the potential problem and stop the ever progressive liberal agendas and let freedom and choice as a principle be allowed to happen and cease the dogma of 'we know best' and desist from forcing it upon the public.

Such bodies, government ministers and media editors need to reflect the decent principles we are in danger of losing and not let these latest liberal concepts restrict economic growth, indeed they should support growth.

If they do not and we continue with the narrow restrictive approach and limiting 'freedom' the predictions above will come to pass!

STATE CONTROL – AT WHAT COST AND WHAT FUTURE?

The state is taking over – in the UK and through Europe – in its various guises, with National Governments ever increasing laws and through local Council's and regional Governments and Assemblies. Evermore controlling is the EU organisation adding more legislation and restrictions in every aspect of people's lives.

Freedom as we have known it is diminishing and even freedom of thought and expression is being discouraged. We have seen this with the introduction of "offence" laws, badly drafted, to give the state in the form of Police forces the power to prosecute people for saying different things to the 'accepted' politically correct status.

The pettiness of this move of the state to dominate is clearly seen by the actions of the Public Sector and local Council's flexing their muscles on pet issues, exemplified by refuse collections and the obsession of demonising the public for having their waste collected under draconian regulations. It used to work well for decades but now it has become an industry on its own, created by the faceless bureaucrats.

This, and other forms of control, is implemented and sold to us on the basis that it is "good" for us and for the environment or other such excuses. The position is that 'we', the authorities "know best", even when we, the public in the main, don't want it and in the face of evidence that it doesn't work! This situation is degrading the relationship between government and the people and in some small way has been evidenced by recent election results.

How did this come about in the UK? We have always had a tendency for officialdom but the stimulus came during the 13 years of Labour Government between 1997 and 2010. They introduced a plethora of new laws and true to left wing approaches wanted to control everything, to engender conformity.

They emboldened the local Councils and with additional funding for such policies gave them the means to impose and implement more control over the public. The very public who pays for it – but don't let that stop the juggernaut of centralism. And to hell with the cost as we have seen public expenditure balloon under Labour, as it always has.

Another clever move in this centralisation was the devolved government concept. Sold on the basis of "localism" but behind it engendered more government and controls. These devolved administrations were dependent on central Government in many ways, funding being critical, so central control was maintained and enhanced. Political disassociation could be claimed but in reality it has enhanced the typical leftist approach of large government. However, breakaway approaches as seen in Scotland with the SNP has created disunity and is now causing disparity and confusion at an unaffordable cost. The intention, by Labour, to get Scots to vote for them (as a reason for doing it) has failed and with damaging consequences.

We have seen the conditions and the effect of leftist mantra with communism in its extreme form in the USSR and even the fascist element in Germany with the same central, all controlling governments and the failings of these. Arguably there is great similarity of extreme communism and fascism and the effect on the people they control. It is always detrimental in the end.

We haven't got the extreme form of central control yet in the UK but we are moving that way. It is not something the majority of UK citizens want. But governments of all colours have a tendency to centralise. In today's terrorist feared world we are constantly bombarded with security issues which give government agencies wider powers. The threat of climate change, whether real or not, is used to give collective focus for control, control across wide aspects of our lives such as energy, transport and, of course, taxes.

What then does the future hold? In the extreme our society would suffer dominance similar to Stalin's communism in an Orwellian structure with "Big Brother" control and we all bear it. More likely as such moves of dominance grow, bit by bit, the 'public' would, at some point, rebel either by organised means and within whatever political system is available or by revolution with violence that this provokes. Either way at some point there is likely to be a hiatus to combat the centralisation.

The catalyst for this is likely to be economic in that the cost of government and its machinery will become unaffordable and the impact on the public at large will be unacceptable. The other requirement is a leader of a body to identify with objection to such centralism and controls and offer better and more acceptable alternatives. The timing of this is unpredictable at present but at some point within the next 50 years is likely. The tipping point will be a perfect storm of world economic change – likely favouring Asia – and decline of Europe (and our living standards), military conflict around the Middle East and parts of Asia and the USA's intolerance of other countries and their trading demands.

We, in the UK and Europe, cannot remain isolationist and tensions between European countries will grow as the centralisation moves become greater and the EU will become increasingly unacceptable to the public.

People who purport to govern need the tacit support of those they govern, at least in the long term as any short term oppressive control will be just that, short term (although in some regimes it has previously lasted many years) but in today's world long term oppression is less likely. The willingness of being governed requires common sense and fairness to prevail and be seen to be adopted so the nonsense of current issues being forced on the public against their will by the administrators of government, local or otherwise, will need to change. Ignore history at your peril – it's not the low paid pitch fork waving mobs who start revolutions but the intelligentsia and middle classes.

To simply force people to swap freedom for claims of security or safety will not be sufficient long term. Such threats need to be apparent and human nature for freedom will eventually be revived.

To head off this uncomfortable and possibly disastrous scenario from becoming reality, the approach of control and conformity needs to be dropped and governments need to change their stance. A society needs rules and structure but these have to be accepted by the public or the very nature of society will be put to challenge. It may be that the chaos event needs to be met before a more balanced governmental relationship with the very public they represent can be achieved.

The recent trend for governmental control and regulations exacerbated by Covid restrictions is creating a society becoming dependant on such governmental controls. Children are being conditioned to obey and it is blunting original thought and self-reliance. They are fed on central control and depending on it. This is a dangerous situation.

Michael J Cole

STATE CONTROL – FOLLOW UP

The requirements for funding the 'monster' state in all its forms are becoming unaffordable and the management of it is now so inefficient the state cannot fulfil its responsibilities effectively.

The administration of state has become so complex and is becoming so wide reaching over all aspects of people's lives and controlling. It is so wasteful in the extreme and is getting worse. Soon it will become so constipated it will cease to function to any reasonable degree. It cannot manage state support for benefits properly, delays and cut backs are caused by over bureaucracy and the obsession to create more 'rules'. It would rather spend £20 to save £1.

Looking at the burden it is claimed by 'official' figures in 2017 to have reached an average (of those paying taxes) of around 35% of income – the highest since the 1960s. However, add in Inheritance Tax, Purchase and transaction taxes such as Stamp Duty, vehicle taxation etc and for higher tax payers the figure hits well over 50%. Also consider that the top 5% tax payers contribute about 30% of the take it begs the question how long can this balance go on before it breaks or before people rebel.

This situation is evident across the whole machinery of the State, in Government, Public Sector and the Civil Service and all associated bodies. More and more resources and people are paid for from public funds, taxes, to cover the growing monster. The day of more governing than being governed is looming. The demand to feed the monster is getter greater and the burden falls on wealth creators, business and the non-public sector employees.

A question to ask is what benefit is there in this ever growing state for those who pay for it or indeed for what's left of the society it is supposed to represent.

The follow up question is what happens when the state and its demands become greater than the ability of those who pay to actually pay.

The crossover point will be faced when the state demands and the costs to fund it (through tax generation) meet, or at lease gets closer. The negative balance between those governing and in the state pay becomes too much for the wealth creators. Resistance and rebellion will likely become a 'cause celebre'. The trigger point is not yet here but there are signs of it coming.

Unrest in European states is evident and resistance to the EU 'Government' monolith is clear in the UK. Britain's challenges of devolution are already seen and the far left, pushing greater state control and the weak right leaning parties, unable to oppress this effectively will drive the state rollercoaster further along and it is difficult to see any solution to this state momentum. If it does meet challenge it will only come from people rising. This will need a leader or focal entity to collate a movement to promote a peaceful change but this is not evident at present.

The world is also in some turmoil. The Middle East has all sorts of country and religious sector conflict. America has a split personality over Trump's presidency. Russia has real problems over its borders, once protected under USSR organisations but now exposed. China is engaged in its ongoing dichotomy of central control and freedom of enterprise. India and the African continent haven't yet entered this phase so the future is even more uncertain. This is probably the most challenging period the world has ever seen as it faces up to the state control issues against the freedom of people.

State values and its role needs redefining for each of the territories to which each state applies. It needs to define the key principles for which it stands and sets laws to protect these but realistically limited in their extent in order to allow genuine freedom. Without some degree of choice of anarchy real freedom does not exist and we need to become acceptable to this principle and allow good to flourish.

It needs to have a framework to allow the state and its supporting role to balance with entrepreneurial freedom, entrepreneurial freedom of business and society. It cannot be all things and manager every aspect of people's lives (which it appears to be trying to do now) – history is a good lesson in this where it has gone wrong and totalitarianism, whether left or right, has only limited shelf life.

Back to the thrust of the opening position, the issue of change to the rolling state control will be the financial imbalance caused by the unaffordability of the state's demands. This will trigger the resistance by the large populous affected feeling the squeeze and wanting freedom from the state. Better to change now under evolution, to avoid the climax of revolution, rather than persist with the ever hungry monster which in the end would be unsustainable.

Will the state machine and those driving it see the precipice in time to take action?

STATE AND THE MISALIGNED JUDICIARY

Together with the creeping state control and "the patronising disposition of unaccountable power" as proffered by the Right Reverend James Jones KBE in the Hillsborough disaster paper, the situation of policing in the UK is in need of change.

The self-focused approach by police forces and the leadership and the apparent loss of impartiality brings into question the ability of the police in general to exercise fairness and justice. Credibility is now seriously in doubt and the relationship with the public, which is critical in the balance of law enforcement, is in jeopardy.

So many examples of failure and consequential cover ups demonstrate that the faith by the public in honesty and honour is being lost. Misdirection, misguidance and misaligned priorities adds to this wariness by the public about the rationale behind the various operations. There are many examples where resource allocation is misaligned and mistakes happen – not listed here.

Too often police forces hid mistakes or cover up misconduct by putting up barriers to information and "circling the wagons" to protect their own at the expense of public awareness. This is wrong and it cannot be allowed to go on. It breaches the very basis of law enforcement.

The move by Government to use the police to enforce its political demands and engage them in PC pursuits is not only diverting police forces from real crime prevention or investigation but it is using them as a function of Government, to control public views as in the case of the Hate Crime Bill. This was never the role of the police and should not be – in fact their independence is vital if integrity of law enforcement is to be maintained.

A charter of operation and standards needs to be rewritten to commit compliance to openness and accountability to the public and the duties of law enforcement. This should include admission of failures and acceptance of responsibility.

For effective future policing and to regain and maintain public confidence this is a minimum requirement. Moreover, the execution of such a charter is essential and needs not only to be done but to be seen to be done. And the senior officers need to show leadership in this regard and to avoid the arrogance and superiority too often displayed, especially in the face of failure.

The increasing cost of providing public services including policing will, as outlined previously, become even more under scrutiny and given the resistance of the ever demanding state control (and policing is at the forefront of extension of laws) this will become the subject of public resistance if it is not provided with the sensitive public engagement that it deserves and is needed.

STATE CONTROL AND POWER – THE IMBALANCE AND DAMAGE

Additional to and consequential to the state's creeping control is the growing arrogance and supremacy displayed towards the very people the state institutions are supposed to represent.

This now appears to be at both institutional level and individual level by the state employees.

The Bishop of Liverpool, the Right Reverend James Jones KBE, accurately and incisively covered this in his investigation into the Hillsborough disaster, pointing to "the patronising disposition of unaccountable power".

This misalignment of purpose and conduct of state authorities is now endemic and runs through all the state managed institutions including the Police, Councils, Government and its departments and even the NHS. This attitude of superiority also infects appointed agents by state bodies, even privately owned ones – they take an indifferent air to the public once they become appointed. It can be called the "high-viz jacket syndrome" – give people any form of authority and dress them up and they will likely abuse the power.

Why this happens is probably the basis of extensive research. To stop it happening is more to the point.

A new code of operation is needed for the state and its institutions to engage with the public to set a clear set of principles and standards to get the balance to reflect that these institutions are there to serve the public and help improve lives and not create misery, to make things easier and effective not hinder them.

James Jones suggests a charter to outline the key principles and the type of relationship needed that institutions sign up to. This would also cover the acceptance of responsibility when they fail or things go wrong and become fully accountable. This is a good starting point but it also needs an attitude shift by Government to put the public first and control the way it carries out its duties.

Too often we see cover ups, the state bodies using public money to defend their actions and put the public, the individuals affected, at a disadvantage. Too often we see the bureaucratic process, invented to protect the state, put up to obfuscate the process, fairness or the exposure of truth. Even small issues (car parking and refuse collections come to mind) are shielded by state supported administrative blockage to dominate the individual.

Under these conditions reason and justice are often sacrificed. The public feel cheated and angry.

Put this alongside the creeping state control in all its guises, the unaffordable cost of government and it breeds discontent. The frustration and dissatisfaction will grow – it is evident it is happening now – built up particularly over the last 20 years in the UK – and will possibly result in real public challenge.

Couple this will the potential unrest and financial constraint, Government leadership needs to do just that – lead and set a change of tone for the relationship between state and its public.

OUR POLITICAL STRUCTURE IN MELTDOWN

This paper was written before the Brexit deal was finalised – but the points made about Parliament's approach remains valid as a critique of its behaviour.

What right has any of our MPs to avoid carrying out the legislative will of the public and try to thwart Brexit? We were offered a vote and promised our wishes would be delivered. If this does not happen then democracy, as we used to define it, will have been abandoned.

The public, whether for or against Brexit, will see this as a breakdown of what we have considered fundamental to British standards. The world will revise its respect for us and we will be castigated and ridiculed as a hypocritical state. What is the difference between a despotic regime that ignores its people and our Parliament when it behaves in the same way sacrificing any integrity for the elitist's will.

We have held elections for generations and supported the elected party, even though often they were only successful on around 25% of the electorate capacity and we never asked for a re-run, so why, when the Brexit Referendum was such a clear win, should the Parliamentary losers conspire to have it overturned? Why are the so called 'honourable' members doing their best to manipulate the 'system' and rules of the process to interfere with the public's will?

Bercow and Grieve who have apparently been exposed in a disgraceful pact to engineer processes to stop Brexit should be ashamed. They should be brought to book and the public needs to stand against such treachery.

If Brexit is thwarted it would be catastrophic for our form of democracy and it is likely to damage irreparably the trust in Parliament and our system. It could, and likely, result in civil unrest both directly and indirectly. What faith and respect could the electorate have in our system of government and the people representing us if they can cast aside a legitimate referendum result and renege on the promise of delivering it just because they don't like it?

The prognosis is dire. It will eat away at the balance of the relationship between those governing and those governed and likely stain the whole fabric of our social and economic status. What faith would people have in our structure of government if the collection of MPs can simply play politics and peddle their own person and party agendas at the expense of what is right for our Country. To allow the standards of behaviour we expect from our political system to be subverted and degraded in this way is to hit at the very heart of what Britain stood for and the reputation it guarded for 100s of years.

There are many examples of the public standing up to the ruling classes, the Magna Carta is clearly one, and it is possible that this issue will cause a fundamental challenge and possible breakdown of trust if Brexit is allowed to be overturned.

The 'May' deal as it is described is not totally what we would like but as in business negotiations sometimes you agree to a stage and then negotiate to close later. This is a case in point, we can deal with other elements before the commercial deal is finalised. So we should take her deal.

The way negotiations have been handled is amateurish - to agree a payment before you have a deal was naïve and to have interference from an assortment of 'onlooker' MPs during negotiations is ridiculous – imagine a corporate entity doing such a thing. But we are where we are and to have the Remain campaigners in Parliament blatantly changing the rules to stop the process is unacceptable.

Those doing so should think hard before voting down the deal because the public will not accept it if this means Brexit is lost – at least without repercussions – and it will degrade our society and political structure for years to come if Brexit does not complete. Such moves are foolish and irresponsible and demonstrates that the political system is not fit for purpose – it has been declining for some time and this situation has exposed its faults. Much damage has already been done but to revoke Brexit will tip it over the edge.

Can we please stop the political game playing and can those protagonists of mayhem revise their position and think of Britain. Those of us running companies to create wealth, employment and pay taxes need the reassurance that our elected politicians can manage an orderly exit to deliver the public's wishes and maintain our social and economic welfare and also a degree of dignity.

Michael J Cole

A LIBERAL SOCIETY?

Where will liberalism or a media version of it, lead society?

LIBERALISM

What is meant by this term and how is it applied to society?

It is a general term implying 'freedom' but its application is conditioned according to the function. For example, liberal economic policy means free trade and free markets, promoting entrepreneurism but does not necessarily mean liberal social policy of uncontrolled immigration or abolition of conservatism.

The term has become abused and misused so it is essential to apply it by description to the issue at hand.

The opposites are often true where those supporting liberalism in social issues are against it in economics and vice versa. The media pick and choose its position and the BBC, in particular, take social (elitist) liberalism as a done deal but rail against freedom in economic trading. Their philosophy is "settled" (as so often they apply to any issue about which they have a fixed view) so nothing is offered to balance the argument – in fact no argument takes place on its medium.

This is a key problem, when the media (effectively the BBC in the UK) promote liberal socialism in its coverage so reinforcing a view and providing a platform of propaganda, then this is applied to a whole raft of other issues and liberalism loses the flexibility to apply meaning to different issues. It has become a distorted word.

SOCIAL HARMONY

A perennial question being faced since human social groups were formed is how do you keep all people happy?

For society to function in a cooperative state a balance of all social levels needs to be formed. When it goes wrong there is unrest, war and violence, in the extreme. Revolutions around the world over history demonstrates this. As social structures develop the inequalities are settled less violently.

This issue has always been related to standards of living, relevant to the times, allowing for material and technological conditions, between those at the top, the rulers and those at the bottom, the proles. The ones living in luxury and those struggling to survive or condemned to a life of hard toil.

An organised society can deal with the various sub levels between the extremes and allow upgrading from lower to upper levels. Meritocracy is the foundation of such advancement. At the other end when advancement is not possible a degree of satisfaction must be attained to avoid the break out of revolt.

Revolt needs direction and numbers. The direction almost always comes from the intelligentsia, educated but dissatisfied and motivated. The French Revolution and Russia with Lenin are clear examples. They feed on public dissatisfaction and economic stresses to fuel collective angst to create a movement. The prole masses provide the numbers and the muscle. A working partnership between the thinkers and the doers.

These movements created change and there are many examples across all countries and regions of the world.

A more developed society is less prone to violent change, revolution, unless things become really bad in terms of living standards or trigger events. It is more likely evolutionary change would take place by agreement and through a legal related process.

However, the trigger could be a significant event. It could cause disadvantage to a large enough group to stimulate rapid action and change and even developed societies can face this. Again, many examples are to hand. In the past what might have been considered stable structures have tipped over into civil unrest and change. The Roman Empire is a classic case. Jarrow marches in the UK, US civil wars, Iranian overthrow of the Shah and even Hitler's rise to power demonstrates this point. Also, of course, the decline of the old Soviet state is a major point which arose in the late 20th century.

Sometimes it is a build-up of issues, often economic and sometimes a particular event to trip the wire! It usually is accompanied by a large proportion of society being neglected or enduring hardship and injustice. The joint effect of substandard living conditions and basic needs of survival and the human instinct of justice is a powerful combination to rouse motivation.

In today's societies in so called civilised and developed countries where basic standards of living are considered reasonably good for all, the philosophical issues can also play a part. The way society is managed or policed and what society stands for and who represents it or how it is structured can become a key point of ignition. Controlling governments are not as acceptable as they were 50 years ago and freedom is not always as defined by the liberal elite. The views expressed for unlimited immigration and totalitarianist approach on issues by elitist groups and politicians will be challenged. Ease of communication and social media have enhanced the ability and availability for people and groups to interact and join forces to help create a movement and coordinate challenge.

Such challenges will surface, helped by the electronic communications medium available to all, especially when such policies restrict freedoms of expression and movement and if/when they interfere or restrict the standards of living people have come to expect. Energy is a key component to quality of life and as an example if this were to be in short supply it would become a trigger point and government policy would be balanced by giving likely rise to rebellious challenge.

Plato's 'Republic' in summary provided a rationale for the rulers and the proles in that they have basic inherent value differences which determined their status or standings and that this should be accepted. The proles would, therefore, need to accept this position and so avoid unrest – a paraphrased analysis but here to make a point. This acceptance in the 21st century no longer applies, if it ever did. This is the continuing basic conundrum to face; how to keep social harmony?

The balance of social harmony is a critical issue in social structure and peace. It is no longer accepted that rulers rule and live well whilst the proles suffer. The differences will remain, albeit less fixed or stark, but the basic lot of the non-rulers needs to be satisfactory to the degree of comfortable life standards, freedom and choice and the opportunity to move up the scales or even in the developed world unrest could still emerge.

THE MEDIA (AND THE BBC) IN SUMMARY

The media, and the BBC in particular, are grappling with two opposing movements – the populism, the force of the people whose views reflect the main thrust of public opinion and elitism, the force of liberalists (and the BBC) pressing for their version of how things should be. This is and will increasingly become a battleground. Populism will win, or should. After all it is the people who pay for the effect of government, laws and, of course, the BBC. The isolated centralist BBC will need to undergo organisational or philosophical change to identify with the public instead of trying to set the agenda.

The BBC, the main media forum in the UK and to some degree worldwide, has transformed into a propagandist organisation (which also provides programmes). Its views presentation is laced with bias and angled to promote its own version of events and are subsequently distorted in order to do so. Its programme content is often also directed in a similar way. It is steeped in left wing supporting politics and this influences its whole approach. It engages in twisted tactics to promote its pet causes and denigrate the ones it opposes or doesn't fit into its narrative.

Moreover, the public perceptions and views need to be recognised, after all, it is the ordinary people living in societies who face the daily challenges of living, working and surviving who have to meet the consequences of government policies so their opinions matter more than the liberal elitist views of those in privileged positions or in the case of the BBC arguably out of touch theorists.

In effect it is the BBC against the public, the very people who pay for it. This will form the battleground in future times – if the Government won't or can't sort it then public pressure may well be needed.

Past employee journalists have been critical of the BBC's biased approach. David Sedgwick and Robin Aitken have written detailed books as critiques of this.

Such bias, political and PC manipulation by a public funded media body, using its influence for propaganda of its own narrative, cannot be allowed to go on – it must change. It should be a neutral, objective news media organisation reflecting truthful representation as indeed its Charter requires. Management of the organisation needs to change too in order to reset its approach to fit its Charter accordingly.

Michael J Cole

THE CONTROLLING MIND, AS MANAGED BY THE BBC

The BBC as the UK's tax funded media organisation practices thought manipulation as efficiently as any controlling governing body. Its brainwashing power is immense and David Sedgwick's book, "BBC: Brainwashing Britain?" is an excellent critique and created from an internal observer's position. His assessment is disturbing, frightening and most certainly depressing. Taken in the round it confirms the need to have the BBC change its direction and its management.

The creeping drip feed approach is now seen for what the BBC is, what it has become. It has much departed from its original purpose and from its Charter to which it is supposed to follow.

The roots in the conscious subversive manipulation approach can be seen in the collective terms of Marxism. The destruction of Western values, economic principles and social balance is now underway and in a manner that the public at large cannot appreciate as the concepts promoted are now, after 50 years or more of infiltration, engrained in its whole approach and its version of 'accepted behaviour'. This has built up and expanded before anyone has fully seen it in the round.

The anti-issues approach have been promoted and formed part of its deliberate propaganda, in summary being anti-UK, anti-USA, anti-Tory and anti-West. This is the cornerstone of its Marxist philosophy. No support is given to the opposing views of the BBC's stance on issues such as immigration, gender rebalancing for statistics over talent, climate change and left leaning policies and no opportunity is given to challenge the BBC's narrative on such matters.

On broadcasts and programmes which do touch on these topics they are constructed to disadvantage anyone trying to put forward opposing views and this includes the editing of news reports. Many examples can be seen where the BBC distorts the coverage to give a false picture in favour of its own position. In the case of a news story the position and report is decided before the investigation so fitting the 'facts' to suit their version. Real journalism is sacrificed at the altar of BBC fiction – the 'right-thinkers' need to reinforce their version irrespective of truth or fairness.

The 'wrong-thinkers' are those opposing the BBC's anti-views and the 'right-thinkers' are those fitting the BBC's own narrative of how life should be lived and how the public should behave.

The identity of the right-thinkers are the liberal elitists, to put a title to them, who coordinate their views on topics mentioned above. They consider themselves civilised, intellectual and superior. The wrong-thinkers are the large sections of the public holding, say, traditional views who the right-thinkers think are 'poor, stupid and racist' and are 'proles' not able to understand their higher morality. All this, of course, is a total misconception by the BBC and they misjudge the public and the contribution and intelligence the vast majority provide to the well-being of our society and the right they have to the debate.

Many institutions and opinion forming bodies, including Universities, the UN, governments and even judiciaries have fallen for this rhetoric and so influence has been gained which is then reinforced by further BBC propaganda. A well tracked method used by regimes in the past to incite people to follow.

The BBC in particular, of all the media, sets the presentation and often the content of news and programmes to fit their view of "life standards" and engage in the distorted propaganda to reinforce it. It no longer reflects public opinion of those not being right-thinkers or reports objectively as it sees its role as directing the news to meet its own version of it. This is a most disturbing element of its conduct – many examples can be seen where they have deliberately presented a story to mislead and used numerous tricks to deceive the rationale behind it.

This institution will not change of its own volition, its staff to a member have bought into its narrative and its agenda is so instilled into them they automatically follow the modus operandi. It is an Orwellian Truth Ministry. An appropriate time here to quote from *1984: "The essential act of the Party is to use conscious deception while retaining the firmness of principle that goes with complete honesty."*

The question is what can be done to stop this brainwashing campaign and return it to the principles of its Charter that was established when it was formed? And which Government, who have the authority to do so, will take this on?

A Virus in Society

EQUALITY - OF WHAT?

There is much talked about equality, used by politicians to emphasise their support for the under privileged and to show their caring credentials. But what does this mean.

Equality does not mean sameness. As humans we are not the same – not physically, not emotionally, not anything. We are each unique in just about everything. Similar yes, but not the same.

There will never be equality in the sense of total equivalence – unless we make robots on a production line. As biological specimens with the complicated structure and cellular configuration every one of the billions of individuals will be different in some way.

There will be behavioural traits common to our species, as indeed for every biological being of animals caused by inherent biological similarities and, of course, nurture, the way we are bought up in the local society in which we live (at least for the informative years of development). These balancing factors create the "attitude" of the person and how they react to circumstances throughout their lives.

Yes, their reactions will change and adapt as their experience grows but will be conditioned by the basic biological dynamics and their particular experiences and upbringing.

That said, it emphasises that equality is not, therefore, sameness and treating people the same under the guise of equality is inappropriate.

Having a condition of rules under which people should behave is to protect society and dependants against disadvantages but this should not be confused with the argument of sameness.

Equality, if it has a place, can be defined as equality of opportunity. To have no prejudice on an artificial basis but make judgement on abilities or application – a kind of meritocracy of talent maybe.

Each person will have differences to apply, some more dextrous, some more cerebral, some caring and some not. The physical range is clear – a heavy weight boxer against a disabled person – compare Mohammed Ali and Stephen Hawking, both human but entirely different, but sharing similarities and each making a contribution to life and society in different ways according to their talent or abilities.

They both have, as all individuals should have, the right to exploit their talents and abilities as best they can.

The requirements of a structured society vary and so each person should have the chance to apply their individual talent as they wish and as can be utilised.

The creation of equality is just that, creating a framework for the differences to be nurtured, to make the best of the biological building blocks by letting the environment of development in society, local and general, encourage each and every person to apply themselves.

Fixing the same pay scales or reward levels by government or agents of the state is not the answer, in fact it is inappropriate and counterproductive. It inhibits development and enterprise, individual enterprise and makes people dependant on the state. Loosening of such constraints is required. If some individuals choose not to take advantage of the opportunities or reject the principle so be it but this should not cause governments to impose regulation to create sameness for the sake of it claiming equality.

So, in summary, equality is not sameness and it should not be legislated for, equality is offering an equal opportunity and allowing the difference in each of us to be developed as best it can be.

Michael J Cole

THE PHILOSOPHY AND BALANCE OF RELIGION AND SOCIAL STRUCTURE (AND SOCIAL HARMONY)

A fundamental question is how to do these issues meet or work together?

In the past when religious dogma set the whole standards of living and of human behaviour it was clear. In some religions today where their dogma or tenets are laid down it is clear and if a country is dominated by a particular religion it can be claimed or seen to have an obvious influence on social structure. In mixed or 'developed' countries or certain geographical areas it is not so clear. Nor does it lead to harmony, in fact it often leads to disharmony and conflict. Often this also leads to persecution of one section of society or another. In earlier centuries it caused wars and examples can still be seen in today's world.

Unless there is a common agreement of only one religion or set of principles for behaviour then there will always be the prospect of disagreement and in the extreme violence as the beliefs are so engrained in the strong believers that it becomes a 'crusade' to invoke those beliefs on others.

The question then moves on to how can a balance be struck to let each and every religion co-exist with others of different persuasions?

Within a regulated and tolerant society this can be accommodated if all parties agree and look at each other to see the 'good' principles of their respective religion or faith. For less regulated or developed societies this is clearly difficult or impossible.

Male and female disharmony and the feminist movement over the past 100 years started by specific campaigns, like votes for women in the 1900's which then evolved into a structured campaign on a number of fronts and issues and extended through women's contribution in the second world war in particular in the 1940s and then into the 60s when emancipation accelerated. In the late 20th Century it became embraced in laws and behaviour. It has largely been successful as women's rights in Western societies have been made equivalent to men. Some would argue there is much still to be done but by and large the argument has been won.

Homosexuality has been another issue which in recent times has been settled in terms of social structure and rights in the Western world. Attitudes with some people may still be biased towards such orientation but again the argument has been won.

Those religions not accepting 'equality' of women or homosexuals are not inclined to accept the principles adopted by Western societies and along with some other fundamental issues it will continue to create major points of conflict.

How to change each viewpoint to bring harmony is the nub of the problem. Western society will not compromise on the hard won position achieved over long periods. The fundamentalists not accepting the position will likely continue to resist adopting Western views so harmony will not happen (at least not soon).

Change happens slowly in such matters. The religions or dogma not accepting 'equalities' of this type will take time to change, even if it is possible. Indeed the Western, Christian world had prejudices of their own in these areas which eventually changed, so it will take time. Saudi Arabia is bending to a degree now with small steps to let women play a part in their culture which previously was restrictive.

The solution is not clearly defined. What will help is trade and standards required to do business for the motive of prosperity. History shows this is a real generator of change. Isolation is not the answer as it reinforces self-justification of regimes and their leaders. Like human rights, trading needs and opportunities is the catalyst to bring about fairness and 'equality' to promote the harmony often and usually denied by religious restriction.

Social harmony is the culmination of many small steps of acceptance by groups, possibly by having common aims or goals and having a mechanism to allow understanding and acceptance to be cultivated. Trade is one such mechanism. The main trading world countries have a chance to influence this but there needs to be constructive cooperation on both sides of an agreement. There will not be a common pace across all regimes as each will face a variation of philosophical movement towards a common or joint operating agreement.

The standards of Western values are considered as fair and equitable by the West, other cultures don't necessarily agree and bridging the gap is the challenge. The question here is how each party will accept the compromises and to what final level is the end position.

Much is talked about multicultural societies and trumpeted by some political groups and the media, including enthusiastically by the BBC, but is this a series of isolated cultures within the total society (or country) carrying on separate from the main society and related laws, whereby the main society only tolerates the isolated group? Or is it to be a concept of complete integration and conversion of the incoming culture to the main society, so losing any selected identity?

A Virus in Society

This is the fundamental issue to be considered and tackled. A tolerance is possible but the laws of the main society need to be clear so equality of dispensation can be administered and seen to be so.

Acceptance and tolerance of differences between, let's call them incumbent, societies within defined countries or territories, is possible to a degree but within such areas will always be problematic and if there are issues of fundamental behaviour differences it will be impossible.

Multiculturalism is a false term, it cannot be practiced in full harmony and a 'host' country or territory will prevail. The incumbents will need to adapt or at least accept the laws, practices and behaviour of the society into which they have come. This applies to every incumbent wherever they originate. Tolerance to have some continuity of their base beliefs will be at the discretion of the host and practices will need to comply with the host's laws.

The Middle East, Africa and Asia highlight these difficulties where mixed religions exist in one country and are often shown in violent and extreme situations. To a lesser degree Western countries can have examples of unrest, usually confined to local areas. There is then the issue of differences between adjacent countries or territories where religions and cultural differences mean cross-border conflicts. Particular 'crusades' can be seen by extremists crossing countries to inflict damage and atrocities to demonstrate their challenges.

Much of this is engrained in an historical backlog, eg Muslims and Jews, Hindus and Muslims, Sikhs and Hindus, Christians and Romans and, of course, Catholics and Protestants. None will be settled quickly or simply. The common factor which can play a part, as a concept, is trade – doing business together to make the lot of the people better, to help provide good basic standards of living, education, opportunity and equality of opportunity.

This is not the panacea for solving all the problems, and it would need to be managed carefully, but it is, at least, a focus. Maybe over time it could help stop the separation and pull the differences together.

A REAL VIRUS!

Given the COVID-19 or coronavirus situation, infecting the world in early 2020, a comment on this is probably apt in relation to the title of this book.

The actual virus, the biological virus, is clearly of concern especially in its virulent expansion across the globe. In modern times of international travel such spreading is explainable. The world has seen such epidemics before with varying degrees of seriousness. From the Middle Ages to the present day transmission of viral infections by human to human contact can be tracked. In most cases the virus either dies out or as in more developed times a vaccination is developed and an immunisation programme is coordinated.

The issue about the COVID-19 virus is the reaction by governments rather than the virus itself. Let's consider this further.

Viewed objectively the reaction mirrors a kind of Hollywood disaster movie with widespread panic followed by a salvation of an anti-body to save the world. There, the theme of saving the world rears its head again to justify draconian measures for the "good of people" and as we have all been told repeatedly "to save lives".

Firstly, examining the numbers is not that clear. How many people actually died caused directly by COVID-19? No absolute number can be quoted. Deaths are recorded as "associated" with COVID-19. It may have triggered existing conditions in victims but another virus could have done the same.

The condition of one country, or local area is not necessarily analogous to another so comparisons are inconsistent and so are conclusions put forward by the 'experts'.

The measurements against normal deaths has also not been made clear or fully analysed so again the quantum of the virus impact is really unknown. No numbers of people actually having had the virus and recovering is known so together with other non-accurate data a fatality ratio cannot be established.

Secondly, and this is the more worrying aspect, are the control conditions put in place. Large proportions of businesses have been stopped operating, a position imposed literally overnight. The impact of forced inertia will have a long term detrimental effect on the world economy and individual companies and especially personal and private owners, likely to devastate many who will not recover. The fragility of the economic balance and financial management has been brought home in a short period of weeks and governments have not really allowed for this.

Governments have offered financial support programmes but the big question is who will really pay for this, how much will it cost and how long and effective will be the recovery and how successful will any recovery be in what is left of the life time of those affected.

Is the cure worse than the virus?

This brings the consideration on to the third point of the effect of the impact on the social structure and well-being of it and of individuals.

The lockdown period or periods as it is called have put restrictions on people's movement, severe restrictions. It has also been given legal standing for people to comply. In totalitarianism regimes it is expected and seen to operate for any cause determined by those who rule. But in Western, free societies this is alien to the principles and standards of a free society.

It is explained away by "for your own good" argument or by "saving lives". This justification appeals to some, maybe many, but it is based on fear and guilt if you do not comply. It was brought in overnight and what is to say that governments can't do this again whenever they decide to on any other 'for your own good' issue.

Lord Sumption and journalists like Peter Hitchens and Dan Hodges have written eloquent articles on the dangers posed by this unilateral control imposed by the UK Government and these comments should act as clear warnings for the future.

The impact of such a course of action is wider than the virus or protection of people, it seriously is at odds with human nature. The spin off actions such as apps on phones to monitor movements is another move towards big brother no matter how it is justified. Individuals and societies need to ask themselves if this is the type of social environment in which they wish to live.

Whilst the COVID-19 virus is serious and devastating for those who have been taken ill or died either as a direct result of as a consequence the whole situation has to be viewed with balance. Humans have faced epidemics before and even a bout of flu can cause death in some. A broader view of the value of a successful economy and the benefit it brings to the quality of life needs to be considered against the temporary problem of a virus transmission to arrive at a balanced and practical approach to dealing with it.

Quality of life is more than just breathing or existing, it is about all the other functions of humanity, interaction and living in a developed economy and the life style comforts this brings.

So, in conclusion, the science in terms of measurement is not clear and policies decided upon have not been fully considered with this in mind. An immediate or even knee jerk reaction was taken largely for political posturing by being seen to be doing something. To unravel these restrictions will be another political decision to justify the lockdown position in order to claim it has worked as it would have been worse had they not done it. The central operation has not really proven to work in curtailing the spread, other countries like Sweden didn't do this, at least to the same extent, and its reports would suggest it fared better than those who took extreme action.

The process of control and police enforcement is a real worry for a free society and a risk of it being repeated. And then the impact of damaging the economy and social well-being has been disregarded in favour of the political will to be seen to having taken action.

The repercussions of this will be long lasting. The economy will eventually recover but for those directly affected by the "shut down" it may not be in their lifetime. The other risk is that the door may be open for governments to invoke control whenever they deem it necessary on other issues, maybe so called climate change next, which will be a real challenge to freedom.

FREEDOM AND LAWS

Laws are basically designed to construct a level and methods of behaviour in a social context, usually for a country or region under an ambit of centralised control. Such laws are to ensure conformity of behaviour to standards largely considered able to let the social community exist and interact to common patterns and to a degree of harmony.

Laws are, of course, set by the law makers whether individuals of influence, groups of governance or by general acceptance depending on the format of the structure. There could be tyrants who would influence them for their own benefit or their version of what they consider essential or even good. They could be religious leaders, as was much of the ancient approaches as guidance to fit with certain beliefs. In modern societies the process of law making is largely done by committees or a form of collective government as practised in Western cultures.

The framework of law making is really the central function, to outline a set of behavioural limits to promote collective harmony. In essence this could be ascribed to principles of fairness, good sense, protection of the vulnerable and to isolate and stop acts of some to disadvantage others. Good and reasonable motives.

However, another key principle surely is to preserve the essence of freedom, that balance exercised by choice of individuals or groups to act or behave as they wish. The limits set by laws by definition, therefore, need to take into account this balance – the restrictions on one hand but protection of freedom on the other. And this is where the challenge is seen.

How to ensure this balance is reached is largely a function of the governing process and the collective requirements and opinion of the populous or society at large. This latter 'view' is not easy to gauge as there will be varying versions in a large society and getting a consensus has many practical difficulties. Government structures and election of officials to speak for or represent the people is one way of doing this, to set the balance.

This can and has worked in some societies for long periods and has been developed and extended over time. Whilst particular, individual laws have been changed or added to over time to reflect the social and technological changes the fundamental criteria of the need to control behaviour and to protect freedom remains.

It is this balance that increasingly comes into conflict. How far can the laws be put in place, enforced and administered before the populous or large enough numbers of it challenge them.

This challenge can be revolution in the extreme for rapid change or evolution by adapting a system to reflect the 'majority' or influential view. If it is one issue it is likely to be the latter approach but if it is a whole raft of restrictive laws encroaching on or eliminating 'freedoms' then a major change will be forced upon the governing body. Many examples in history can be seen to show this.

Governing bodies, therefore, have a responsibility to be always mindful of the need to get this balance right. Too much authority will overstep the balance to the detriment of freedom. Too little or loose law making could encourage more social disruption arguably by minorities disadvantaging others.

The governing body or bodies influencing this balance is worth considering.

A single government structure is easily identifiable as a point of decision making. It can be accountable and responsible for laws made and keeping the balance. Where other controlling bodies come into play it becomes more difficult and more confusing when considering accountability. This can be seen in the USA where there is central government and the state structure. In the UK the devolution of Scotland, Wales and Northern Ireland has added difficult responsibilities and law making parameters to the governance. Europe has seen this dilemma with the EU having law making influences across the 28 (soon to be 27) countries.

The latter case of the EU demonstrates the dissatisfaction that can be created when one central law making body is discordant with a local one. And when this eats into the sense of freedom it triggers conflict. This was clearly a position influencing the UK's populous vote to leave the EU with Brexit. Such a status means total consistency across the whole EU territory and in each jurisdiction is not possible except on specific or basic issues and confusion of acceptable standards and execution of law enforcement exists. Devolvement in the UK causes this as well, made worse by differing political standpoints.

So, we see that when the balance of laws for control oversteps the perception of freedom it becomes unworkable in the extreme and is forced to change by either evolutionary agreement or by revolutionary force.

The future and the position here is that more and more incidents in today's world will see such evaluation of the position of laws for control versus laws protecting freedom. The rapid and easy communication between people across the whole world makes such information and evaluation open to scrutiny and any inferences and imposition upon the freedom principle will be challenged. Classic recent examples are seen in Africa, Middle East and, of course, Asia where restrictions of people by legal structures are under threat. This is of current relevance in the case of Hong Kong as China inflicts new laws on freedom.

The conclusion is that government can no longer impose laws and resulting constraints on society at will and they must also go back to the basic principle of laws to take cognisance of the effect on freedom – not legislate for freedom as this can often have the opposite effect and curtail other freedoms (as we have seen with freedom of expression and hate crime legislation as an example) but to consider a genuine balance of choice of freedom.

Short term emergency controls, set in legal terms, encroaching on freedom may be possible but it needs to be just that, short term. It also needs to be used sparingly. Terrorist attacks have created restrictions for society and some are longer term than others. The COVID-19 worldwide virus epidemic has provided severe restrictions on freedoms. These must not continue beyond absolutely essential periods or extent and they must not create a precedent to invoke such regulations at will whenever a government decides to do it on any particular issue.

There is an ever demanding requirement for government and law making bodies to be aware of the need for minimum laws for only essential control of behaviour for social harmony and the overwhelming need to ensure freedom is implicit in every aspect of such laws.

Can governments resist the temptation to set ever more regulations at the expense of freedom? Those favouring big government and central control will always press for laws to contain and historical examples can be seen where these have been overthrown. Such change is usually caused by the masses who are disadvantaged and often motivated by middle classes objecting to loss of freedom and any resulting financial or social difficulties.

Governments of a more 'open' approach still need to face the challenge of keeping the balance!

Michael J Cole

CRITICAL MARXISM – CONSTRUCTIVE DISMISSIVENESS

The trend in recent times to pitch one part of society against another is expanding. It is Critical Marxism.

What is Critical Marxism? It is the deliberate process of creating differences and exploiting such differences between groups to propagate and promote this to damage society and generate unrest and division. To be critical about anything that can be used to demonstrate division.

This builds up and on a wide number of fronts engenders the impression of disharmony and even anger within a society. It undermines the fabric of collectiveness over a period of time. Have this repeated in the media and it reinforces the disparity within society. After all crisis and calamity sells news.

What is the rationale for this? It is a mechanism to create a feeling of uncertainty, unrest and even fear in society which leads to the feeling of change. This can then be exploited by those initiating such alarm and can lead to persuading people to vote for those offering to put right all that's wrong. Created wrongness rather than actual problems.

To stoke up fear and unrest is a powerful background movement for political change. Left wing and even elitist liberalism thrive on central control. Political correctness and themes like it create a platform for differences or different opinions which as it develops creates unrest. Normal acceptances of different viewpoints by individuals or groups can be accommodated but with 'campaigns' or pressure groups it can be portrayed to create division.

A Virus in Society

Are these movements purely designed to deal with a disadvantaged group or does it have a wider agenda – are they designed to exploit differences to incite greater polarising of positions? By doing so it inflames emotions and calls for government action and laws are often the result. This further causes difficulties and unrest.

Motives of fairness and human rights are tagged onto this for justification and the process gets repeated and expanded. It eats into the normal balance of social behaviour by having antagonists exaggerating the position.

Look at the PC or woke issues now facing societies. The need to conform to behaviour and thinking to these created standards is expanding. Any minor issue can be dragged into the ambit. These are initiated by arguably so called 'liberal' supporters. So called because they are not liberal in that they demand conformity and 'wrong thinkers' are not condoned!

Persecuted groups such as women and homosexuals where intolerance and even laws were in place to disadvantage such groups needed to be resolved to give genuine equality. This has largely been done, at least in the Western world. Race and racism is another clear example of the essential improvements made. And the general movement of equality for all people irrespective of colour, creed, race or disability is a sign of a developed society. This is not the issue raised here. What is the issue is the destructive approach of how the application of such change is manipulated to create the disharmony.

Pitching social groups against each other is a common tactic. Muslims against Christians, transgender against fixed gender, immigrants against indigenous residents, Tories against Labour, Brexiteers against Remainers and so on. Such differences are in themselves a balance of a mixed society but exploitation is the weapon used to create more disharmony. This then leads to possible friction and opens up a platform for political positioning. Even activity by nations conducted hundreds of years ago are used to denigrate a country or say the British Empire when whatever may be considered wrong at that time by today's standards the events are not related to people alive today. But this is used to create anger and disharmony by those pursuing such an agenda.

Is this a coordinated campaign? The answer here is difficult to assess. At one level it is a number of individuals passing a particular agenda, at another level it is a movement, often within a wider group or political party using the communications medium to promote their position. Social media communication makes this much easier now than in earlier times. Such exposure gets momentum and even small or modest groups or concepts can get notoriety and traction – even government involvement and legislation.

This is the problem, a small almost insignificant issue can be escalated to prominence.

The collective impact drip feeds the propaganda to create a feeling of disparity, or even in the extreme, despair. Even good news is tainted with negative inflections. This then is the approach to be critical, to undermine the harmonious social interaction to open up an opportunity to impose control, to fix the 'problems' to solve the unrest and smooth the divisions. Such disharmony is largely a created illusion but it is a tactic used by Marxist regimes and often successfully. Liberal elitism, as a collective name, is actively practising these tactics.

The media bear a responsibility here. Their communication systems can be used to repeat and reinforce a particular notion or

PC position. Organisations have their own political affinities and the BBC, the UK's national broadcaster, has an influential position which they exploit to their own ends to good effect. Its own pre-determined narratives set the agenda and news and programme making are constructed to fit such narratives. It becomes a propaganda medium and drip feeds their chosen positions to promote them and influence, some would say brainwash, the public.

Many examples can be shown where this occurs including topics like climate change where they block out any challenge, their support for the EU and their anti-Trump, anti-USA rhetoric which are all presented to create friction between opposing parties. Books by David Sedgwick, "Brainwashing Britain" and by Robin Aitken, "The Noble Liar" are full of examples where the BBC create disparity.

The question is – can this be recognised and avoided in our modern societies to not allow a creeping distortion of our social balance?

Michael J Cole

MARXISM – CRITICAL THEORY
The approach to create subversive change

The subversive moves to infiltrate institutions of influence and the 'minds' of people is to change, to destroy and restructure society in the 'West' – USA, UK and Europe.

Its aims are not so much what it wants to establish but more what it wants to destroy – to bring down. In its place it wants control, power for itself – its tenets of how people, society should be managed and constructed. It would not/should not be free to think to act or create – capitalism would be excluded, entrepreneurism not allowed as this requires 'freedom'.

The processes of infiltration are the fundamental steps. It is to get the public feeling dissatisfaction with the current status. Coupled with this is the tactic to criticise the current culture and status to make the masses uneasy. Then pitch factions against each other – social classes, culture, sex and religious groups to enforce the differences and discontentment. Add in the PC (politically correct) movement and hate crimes syndrome to restrict independence of view and an environment of unrest is created. With a reason to then control behaviour.

On the back of this, influence in universities and naïve youth can be further manipulated and over time you have institutions full of 'influencers'.

This creeping agenda is used to undermine the stable and economic progressive structures and societies (even with their faults) that the West has used for prosperity growth and advancement.

Other aspects are able to be added on to the totalitarianism of this Critical Theory approach. Human made climate change is a real case in point. It encompasses fear, survival and guilt, as it is humankind threatening our planet. It is a good cause, it does not select any particular group or nation so it is generic. How can anyone challenge it?

But by following this the demands are to change the Wests' (and developing countries) economies, for the worse and the social structures, for the worse and allow the new doctrine to emerge and invoke control.

'Man' is guilty for 'poisoning' and 'destroying' the planet. Western societies are criticised for action 100s of years ago and need to be brought to book – this is already evident in state apologies over centuries old issues.

So, the creation of discontentment, the guilt construction, the effective brainwashing in education and influences through public bodies and policies is the establishment and process of Critical Theory based on a Marxist philosophy subtlety infiltrated into Western society by unelected means. Discontentment opens the door for radical change.

These attacks on our social structure are not open terrorism but a quiet set of strikes on the very fabric of our life and standards that we hold or held as fundamental. This creeping unobtrusive threat goes almost unnoticed and MI5 type surveillance is of no use - it is the vigilance of the softer issues that is needed. Arguably this is more serious at damaging our way of life than terrorism as it is more long lasting and engrained in our very structure.

This process in the year 2020 is already established and well entrenched. It will creep, almost unnoticed unless it is recognised and a challenge mounted to combat it and support freedom of thought, freedom of action, economic prosperity and development and a conscious drive to maintain the West's base of commercialism and social balance. To let this creeping Marxism succeed will result in economic and social reversal and likely deprivation – then there will be real dissatisfaction and likely violent revolution at some stage. The motives of discontentment can work the other way too but the West may suffer decline as a result before its revival!

Already it can be seen that changes are being imposed upon Western cultures and beginning to effect economic performance. These are being driven by the areas targeted by the Critical Theory as evidenced by PC initiatives, multi-gender concepts, climate change propaganda and anti-business criticism. Such changes are being enforced by legislation and a diet of propaganda by the media in general and some in particular such as the BBC.

Politicians are becoming more supportive of these changes by those in education leadership and other public institutions. This can be understood from the position of the Critical Theory approach through education systems and exposure to the creeping philosophies some years ago and those exposed are now in positions of influence. They are implementing the theory through the very foundations of our structure in Government, Judiciary, Civil Service, education and media.

It is important that the public in general and key individuals appreciating the dangers of this can stand up and take action albeit against the weight of the raging Critical Theory philosophy and establishment.

POLITICAL CORRECTNESS GONE MAD (AND NOW WOKE!)

In coordination with other papers this topic overlaps with liberalism, social harmony, equality and the philosophical balance of religions and beliefs.

What has happened and is still happening is the demands of particular groups or sections of society and supported by what can be described as elitist liberals to engage in campaigns to promote the 'requirements' of such groups for recognition or for rights within the framework of society.

These 'requirements' develop into demands for support and laws to enforce acceptance by society as a whole. As justified as these demands might be or seen to be by definition limits objection by those who do not agree with or recognise such requirements. This is the cause of disharmony.

The actual necessity for specific recognition for such 'requirements' and for specific laws is the question – the wider legal framework can be considered sufficient but the culture of political correctness, the need for everyone to conform to such acceptance means that it is not enough.

Take for example the issue of transgender people. Maybe born with one gender but wish to identify as a different gender or even change gender by medical intervention. The law for allowing people to exist in peace and have equality regarding employment without prejudice etc should be sufficient but some claim it is not and specific additional laws need to be put in place.

These additional laws for recognition have a knock on consequence which means further laws need to be adapted as well. This can be considered reasonable and fair. However, some of the consequences are more complicated. The issue of gender neutral toilets or changing rooms is brought into question. A demand by a 13 year old girl to stop her Council from operating these has been passed. There are many other examples of 'Pandora's Box' being opened such as inmates in jails where 'men' identified as females are placed in female environments and some cases of sexual assault have occurred.

The balance of the effect on society of such moves about the examples above and many others of 'PC' is a real challenge. Often the issue is around very small numbers and it is a legitimate question to ask if it is worthy of reacting to every claim for special treatment rather than let the broad thrust of laws apply. Yes, laws must be reviewed and adapted as society develops but the basic principles arguably need to be followed.

The next question is the one created by discrimination. This is no longer centred upon historically disadvantaged sections of society often summarised as sex or racial discrimination. The 'new' level is laced with the 'guilt syndrome'. Let's consider this further.

Western societies were arguably constructed around the white population and organised by men, older men with status and experience taking the lead. Whilst this caricature is changing it nevertheless has an existing perception. So being a white, middle aged male is something to liken to a dinosaur, passed its oppressive sell-by date – no longer PC – guilt ridden and in need of replacement.

This may be an exaggerated assessment but it forms the basis of 'woke', the somewhat vague and flexible description of out of date and largely offensive approach of the past.

Not really true or justifiable but a modern attitude to use to justify change for PC thinking. An example of such guilt ridden feelings is the demand for reparation for invasion of territories hundreds or thousands of years ago by a particular country or peoples against those who operated in a different culture and time. The present inhabitants have no direct relation to that country or peoples living today. But there is emerging a need to 'settle' things by creating such claims.

Another example is seen within the security service in the UK when a senior official embarked on a speech of self-flagellation about gender and racial restrictions in employment in times past and said he was sorry for being white and woke! Such cultural repositioning is a warning sign of PC and woke taking over common sense and real equality where such demarcation is not necessary.

A move by the Labour Government during the extended period of office between 1997 and 2010 was to change the view of such offences of hate crimes and incitement related to disorder and public safety. The move embraced issues of terrorism and abuse through race, gender and other prejudices. This wide ranging bill and legal positioning claimed to be for fair and just reasons, actually fundamentally changed the balance of legal interpretation.

What it allowed and prescribed was that the Act put the claimant, who felt to be abused, as the principal figure so that the respondent, who was accused, had the burden to prove innocence. This is the opposite of any common law where the onus was to prove guilt. Moreover, someone not directly involved with any exchanges could claim they felt abused on behalf of another party. In effect it promoted victimhood, a theme too often at the centre of the PC concept.

This change of approach can be considered a misapplication and unfair in its intention as well as the legal drafting. What it has done is open up the whole attitude to compliance and political correctness and the restriction of free speech. Indeed any comments, joke or view can be challenged by this law. Who is to say whether someone really is offended or whether the person intended to offend, which was the previous test. This has excited the would be offended and has made the tenet of free speech and free thought a questionable principle in the UK, a tradition well respected and fought for over generations.

There are many examples of the police pursuing cases of 'hate crimes' by reacting to a comment or act by someone when no 'hate' was intended. There are even cases of police arresting and interrogating people with requests to explain their thinking! A court case in 2019 was won by one person tackled by Humberside police on such a matter and the court sided with the complainant – at least a glimmer of hope that such action by the police can be restrained. But why should this be necessary?

There is an argument that there is really no case for such laws constructed as they are and certainly no case for the police to be pursuing thought crimes!

This restriction of free-speech and effectively freedom of thought is extremely worrying. Even universities are eliminating speakers from visiting to talk to students if the subject or text is not approved and fits with the organiser's narrative. In other words it is editing the learning process. If universities, the centres of learning, development and expression of thought, cannot be open to such freedoms, what hope is there for society as a whole?

Indeed, managing the exposure of matters to our children is becoming obsessed by any damaging reaction it may cause them, the snowflake generation as named. But this type of sheltering of truth, life and alternative concepts is ultimately damaging and restrictive to both the individuals and society as a whole. Education and indeed life is about judgement from evaluation and experience and if the scope of this is artificially restricted or controlled then people will be diminished as a result. Voltaire's famous quotation; 'I disagree with your view but I defend your right to express it!' – paraphrased version – is appropriate here.

The position of PC and woke is clearly somewhat at odds with freedom of thought, speech and behaviour and before it takes over the legal process of law making, it needs to be constrained. The richness of a culture is about allowing a wide range of views to be expressed, letting people digest and discuss and having an education system to encourage and support such an approach. Balance and acceptability will then follow. Another challenge; can this balance be re-established and can those influencing it ensure that it happens?

Michael J Cole

TRADE AND ECONOMICS

To consider all the elements of society and social structures the cornerstone is the commercial balance and its effect on such structures and the social consequences caused as a result.

Whatever name is given to the type of economic structure it comes to the same principle – trading. Trading is the provision of goods or services for something in exchange.

In times past bartering was used, the swapping of one set of goods for another – the specialists providing offerings, theirs for others. This developed into a currency to allow it to be used as a common exchange and different goods and services were valued to set a price standard. This is the basis of today's trading platform across the globe. Different currencies have an agreed exchange rate at a particular time of the trade to allow it to function with an internationally agreed mechanism.

This is commonly called a capitalist system where 'free' trade is allowed to operate. 'Free' being governed by rules and regulations however. The fundamental process assumes that goods or services are provided by a supplier and willingly sold and willingly bought by another party at an agreed price and terms. The difference between cost to provide and the selling price is, of course, the profit.

The Western world lets this process be conducted by private individuals and companies and the governments generate taxes from the process – taxes on profits and employment effectively gives it funds to spend on the social fabric of the country – arguably for everyone's benefit.

In the communist model the process of trade is usually heavily controlled and influenced by the government or controlling body. Such structures favour government in preference to trading operations. It is claimed that the population gets more benefits than they would if private traders managed the process. However, past examples have shown that such structures are not as beneficial to the masses, that corruption by the authorities is often rife and the actual economy of the enterprise is not that successful.

What is clear is that whoever does the trade the processes of provision of goods and services, manufacture and supply and distribution to buyers, users and consumers is the same. Management of it and profit share may differ but the overall structure is the same. The efficiency with which the process is coordinated is the philosophical challenge.

Whether it is Keynes or Marx they are really dealing with the same requirements only proposing different decisions of the use of surpluses generated and to a degree as to how to go about managing it.

All this is governed by monetary supply either by country or locally and world banking systems. No individual entity can be truly independent of this as inter country trade is dependent on each other. However, the surpluses, tax regimes and fiscal policies, including lending, is in the hands of individual countries. It is such fiscal management that sets the environment for internal social standards.

One thing is clear, in Western countries, where private companies operate, it is essential that companies prosper, create wealth and profit to pay taxes, to employ people who pay taxes in order to create public funds to support all the needs of a society in a structured way.

This principle is often lost on those political bodies and parties bent on destroying this economic system. The attitude of full state control and management is never fully explained and where it has been tried in part or fully it has eventually failed. Failed in that people's standards of living have declined and on a broad scale the country cannot trade effectively or at all with other countries. Many examples can be seen where this has come to pass.

There are many visions of how the capitalist, or let's call it 'open-trading' system, can be operated but the fundamental point is that such trade needs to be successful in terms of wealth creation so taxes can be raised to allocate for social and society's wellbeing and the provision of health services, transport networks and the like. However the management is arranged or whatever name it is given, this fundamental point is required.

There is, of course, the question of how and how much taxes are raised, and, of course, how it is spent.

Supporting infrastructure standards, projects and initiatives is important and especially with regard to helping develop the economy, trading and wealth creation. These include travel, transport and communications. This is a self-supporting cycle. It needs to be consistent with encouraging business, entrepreneurism and the related activities. So, the tax structure, levels and methods need to identify with this. A punitive system will blunt success and high tax rates are a deterrent whereas low tax and high growth can be shown to generate higher tax revenues.

Key spending areas need to be supported. Clearly, health, education and defence can be considered absolute essentials for wellbeing and security. Welfare support for disadvantaged people is clearly another and to a degree law and order.

A Virus in Society

Discretionary spending on culture, entertainment and environment are needed to provide an enjoyment element to living. This balance of business and society is to be encouraged.

At this point it is important to mention the structure of governance. There is an existing and worsening problem in that the burden of governance and the ever expanding infrastructure is becoming or has already become cumbersome, intrusive, unaffordable and inefficient. The cost outweighs the effectiveness and is increasing at a rate which cannot be funded by the tax payers – at least not without reducing the spend on direct benefit to the public.

Take for example the UK. It has a central government in Westminster, voted in by the electorate. On top of this it has separate devolved government bodies or assemblies with overlapping powers in Scotland, Wales and Northern Ireland. More administration, more bureaucracy and more cost – to supplement what is fixed in Westminster and set further parameters over issues they control.

In addition there is the EU. Although the British people voted to leave the UK has endured this rule and law-making body for over 40 years. The EU is wasteful, spend and tax happy with no audit process to contain its extravagance. It imposes regulation upon regulation and duplicates individual countries systems with no real measure of success and continues to claim huge sums of contributions to support its own structure and profligacy. Facilitating trade is one thing which is considered good but to construct a 'united states of Europe' in governance is a step too far.

On top of all this there are Mayors, holding office with large administration structures, spending budgets and charging the public for such a system. Police Commissioners are another form of governance to somehow work with the Police Authorities and Chief Constables. More public cost.

On top of all this is even more governance with County Councils, District Councils, Parish Councils and their various departments within each of them. They all influence policy and the execution of it. From top to bottom there is a whole plethora of levels, people and cost and the consequences upon effective decision making and actions.

Any organisation with complex and diverse structures has problems of ineffective communication, unsure decision making and unclear responsibilities – the governmental systems outlined above are a precise case in point. Severe streamlining is needed.

It is essential that public sector bodies exercise restraint to focus on essential services only and avoid non-essential or peripheral activities, jobs and departments. They should avoid useless and vain glorious projects which provide little or no value. Gender managers and climate change departments are examples of non-essential spending, however noble they believe the cause to be strict control of spending needs to be implemented. There are many examples where local authorities have obsessed about business rates being upheld when there is no business operating from the premises and where the Taliban style parking patrols have helped to devastate town centres and where such Councils have done little or nothing to invest in or reverse the decline. New Council offices have been built but nothing has been done to help wealth creation.

This type of extravagance can be seen across the UK and the Mayoral appointments have added to this and not least in London being the largest Metropolitan Borough. The explosion of cycle lanes has crippled transport in the capital, increasing costs to the public and business and inconvenience. Transport for London (TfL) is short of funds by several billion pounds – bankrupt – which demonstrates the mismanagement of Britain's capital city. The artificial and non-scientific supported attacks on climate change are vanity projects and have a negative effect on wealth creation which has been sacrificed by such projects and political stunts directed by the likes of Sadiq Khan and his stewardship is a classic example of waste, misdirection and political motives over common commercial sense.

Coupled with all this "government" are the vast numbers of advisory groups, committees and so on initiated independently by each and all of these offices of state with little coordination. There needs to be a more effective balance of central governance for overall direction and local application for relevance to local communities and the structure needs to change to make this work.

More Chiefs than Indians comes to mind – not too un-PC, I hope?

This topsy approach really is unacceptable. Public money is often spend in vast amounts on administration rather than the actual needs of the public or society.

This leads to confusion of policies and direction. It also allows political differences to be exhibited in these various entities so no cohesive approach is agreed. Often this leads to wasteful projects and very often no action at all. The public is duped by this and their money wasted. The whole system leads to frustration and dissatisfaction.

It infiltrates all public sector bodies. There is often little accountability in the vague structures where decision making is so dispersed no-one takes any direct responsibility. The job security protection usually means no one is ever brought to book for mistakes and this is a major conflict with the private sector.

The devolution of the UK, initiated by the Labour Government under Tony Blair contributed to this and exacerbated the problem. Maybe it was done for political reasons and expected advantages to the Labour Party but the effect has been to add to the disparity and division of governance throughout the UK and added to the cost problems. In the case of Scotland it backfired badly with the SNP declaring a sort of UDI and initiating an independence referendum. Although this was rejected by the public the angst of those supporting it is rife.

This is a classic example of creating disharmony and not generating 'local' identity, it, in fact, is divisive and ineffective as a governmental structure. The UK is a small geographical island with 60 odd million people and to have all these sloth loaded structures is regressive.

In short this system of governance does not work effectively and does not provide the public with what it needs. It is so unaffordable as the cost is rising so much it will require tax levels too high to pay.

Change and dramatic change is needed but which colour of government will do the pruning?

It is not expected for government to provide everything for people – choice and involvement should be for individual selection. Nor should government direct or manage the commercial process. In an 'open-trading' approach the development of goods and services is done by commercial companies and consumers will decide if it meets their requirements – it is not for governments to direct. Normal competitive dynamics will provide for product and service development in what is market forces.

The balance of spend will vary by the leaning of the government, left, right or in between but it is considered essential that the fundamental 'open-trading' approach is not dismantled.

If an economy is successful, that is wealth creating, it will benefit all members of society. Some will be better off than others but all will gain. As covered separately, the opportunity and equality of opportunity should be preserved for each and everyone. What is not needed is the dependence culture whereby governments create a feeling of need to make any contribution either for oneself or society a decision for government. A support system for those in need, either long term or short term, should be part of the welfare structure which if open and managed efficiently would be acceptable to all and including business by maintaining a social business balance.

The balance again of wealth creation and social impact is made. The question is, given the economic structures in place, can the government create the support for businesses to increase and encourage wealth creation and spend the proceeds wisely, effectively and carefully? And can government put its own house in order to eliminate the waste of over- governance of the public?

Michael J Cole

POLITICS AND SOCIAL IMPACTS
(A Challenge)

Politics have always been the construct around which societies have been influenced, managed and governed. Whether in early groups, social organisations and sophisticated organised structures we have in modern country divisions it is a way humans behave. Some 'work' more effectively than others but are largely only operable by consent of the populous, either by repression or acceptance.

Politics is the statecraft, setting of standards, direction of authority and framework of belief systems for the area and people under its jurisdiction. It is the principles of adherence, the cornerstone of the ethos of the collective view to be held by those under influence. The administration is part of this but only in the execution of actions to ensure compliance.

It is the setting of the 'rules' of compliance that is in need of examination.

Elections in today's world are designed to let the populous decide who and what type of government they want within the choices on offer. This, of course, is subject to the structure of such elections and they can vary from 'open' or 'democratic' formats to 'closed' or restricted choices. The latter being considered corrupt and the former being fair. However, the influences, media management and financial interests can often be applied to distort the electorates understanding, views and opinions. Clearly all kinds of prejudices come into play but this, arguably, is the effect of complicated social structures.

A Virus in Society

In essence each party or individual puts forward its key policies or manifestos and the public vote for the one they prefer. All this sounds reasonable provided there are no untoward influences to sway the votes. Never a fool-proof system. The majority vote, given various systematic adaptions, gets to govern.

Putting this process to one side it is the next stage, once a government is in situ, that needs to be considered further.

It is this stage in politics, the management of government, that things become concerning. The ones in power and the ones opposing then start to play the 'political games', the media management of information, the justifications, the excuses for failures and the 'sound-bite' announcements. This is to keep control, to present their views as better than the opposition and, frankly, to manage public evaluation, expectations and opinions. It has become the management of news not truth. There is a widespread disingenuity about this form of government. In a high level communications and social media world the need to ensure this is controlled is key and arguably at the expense of facts (real facts) or honesty. Fake news is easy to construct and transmit.

The solely political exchanges between parties are overriding any solution to a problem or issue and largely ignore the requirements of the public. In the UK and the USA in particular the rhetoric has descended into pointless slanging matches with no real thought of the broad answers for the public good. Differences between parties are, of course, part of the play and passionate and even aggressive exchanges are expected and maybe encouraged to form legitimate challenges but this regularly overspills into entrenched political stand offs even when a particular position is shown to be untenable or unwanted. The whole approach has become unedifying and unnecessary.

This is the state of so called democratic politics in the Western arena. Democratic principles, of which there are many versions and systems, is not absolute and will never be totally balanced. It will almost always be the largest group and the one wielding power who wins the nominal victory and it could be argued this is consistent with the democratic result. However, minority groups in the PC controlled world get a disproportionate influence and this often upsets the majority in this democratic maze. Balance and acceptance with freedom in the laws are needed, never to promote total harmony but a better bet than artificial unilateral treatment of all which will never be achieved.

The political dog fighting, however, is destructive and to resolve this politicians will need to adopt a less all-consuming and controlling focus on the political organisation and embrace the need to let the public have their freedoms to interpret their lives and actions to suit themselves and not adhere to the political elitist agenda. Politicians are in a job to serve the public not the public to serve them. A lesson they all need to learn – and quickly!

Too often the management of news is politically constructed and driven. And this is the fundamental point here – media management is greater than the truth – this means politics overrides any honest interaction with the public. Politics become the whole motive of government at any expense of policies for the public good!

No party or politician will admit this but politics is the story not the policies. Many examples can be seen where the failures are hidden behind political shields to the disadvantage of the public. Those involved in government and politics often sacrifice their principles for party unity or political motives no matter how good their principles were they will abandon them to fit the party model.

A Virus in Society

As a result the systems originally designed to reflect the public's views and engage with them to create reflective or representative government of the people has morphed into political machines to propagate their own platform. This is further exacerbated by those politicians driving their own careers and for their desire for power and influence.

In other words the machinery of government is a cause of its own irrespective of the populous being affected. Fine words to say they care and to persuade voters to vote for them are announced but this is for the expediency of power and once in power the political needs take over.

This is a somewhat cynical and even depressing assessment but nevertheless it has merit. In the Western world the established and acclaimed demographic processes are all infected with political systems moving further away from the electorate. The EU, run by non-elected bureaucrats, is a classic example. The rise of the so-called popularism, ie people challenging the elitist or establishment is a sign that the people can see through the dishonesty of the structure and the Brexit vote for Britain to leave the EU and the election of Donald Trump in America are tacit examples of this.

A greater connection with governed to governance is needed. The 'monarchy' of political parties and those in charge of them (or directing or funding them) needs to be broken.

Even the Western governmental structures built on democratic principles are now changing, showing signs of greater state control, of restricting freedom (once admired) and being the lantern of freedom and choice for the alternative of central management. This is being demonstrated by political parties of all colours. This is the real issue facing such nations – do they adapt, for example, Chinese style control, to compete economically or retain open social and economic structures. The people must be allowed to choose. At the moment in the early 21st century the political system, supported by the establishment is moving to shut down freedoms, to preserve their own power base and the public is considered just a necessary casualty of this approach. Is this a case for 'small' government – in approach and size?

As mentioned in other parts of this book it is likely an ignition point of challenge will come when the people, used to choice, will resist the creeping control regimes. This will be accelerated when the economic downturn hits and continues. The political media management will no longer wash and change will be demanded – unless, of course, politicians accept that representative government is what people want and they respond themselves in advance of forced change.

STATE CONTROL

The creeping power, laws and restrictions are being imposed by governments, of once free states on which society was based, which is blocking freedoms and promoting deception and corruption of those in power – and all in the name of "National security" or "saving lives" and often such false justifications.

In reality it is the cover up of incompetence, embarrassment and in the extreme downright fraud. Fraud on a massive scale and misuse of public funds.

The state machinery is becoming all embracing, all powerful and consuming of every part of the public's lives and with criminal intent using bullying tactics.

The Orwellian approach is dominating everyone's lives using any excuse to justify the very practices of totalitarianism of control and against the very principle wars were fought in the 20th Century. The use of media and public sector organisations promote the propaganda to brainwash the public. There needs to be a trigger point to stop and reverse the process. This point is looming and the sooner the better!

Michael J Cole

MIXED MESSAGES

The internet and related social media initially gave the impression of opening up communications. The benefits being exposing oppression in parts of the world which traditionally went undetected until maybe decades later. This was good. It also aided human contact and the good that this offered.

However, there is the opposite side, the corollary. It has opened up secret communications for terrorist organisations to coordinate activities, to allow trolls to hound others and anonymously operate as "bedroom warriors" to persecute others and promote ideas. Often these give the impression of being widespread but usually they are not. It gives a platform for offensive treatment of others and some, in fact now many, are picked up by the media and politicians (eager for popularity) and progressed – even to law! They are often over exaggerated appearing to have support way beyond the reality.

This disproportionate focus is not only unrepresentative but largely damaging. Reason and science is lost. Arguments are made in 140 characters but this is not reasonable with regard to complex issues or indeed scientific ones. It is easy for users to repeat a headline notion but not really understand the topic or the substance behind it.

In fact the opportunities for ideas and freedom of expression have begun to be curtailed as a result – the very idea of expansion of communication has now been lost by the hate crime laws and those feeling offended. It has become counter-productive. Those challenging a particular stance object and those promoting one won't take criticism – people now get "cancelled".

A Virus in Society

It really is Catch-22. The laws of inciting hate crime are too wide and inoperative in a so-called free state and those PC and woke supporters in the extreme seem incapable of dealing with challenge. This has spilled into Universities where anti-woke concepts are not allowed to be debated. These are the very establishments where such open debate should happen – it used to, but the dumbing down is gaining momentum.

There is now a real dilemma about freedom of speech, expression and ideas and the restrictive constraints (aided and abetted by misguided laws and policy) make it impossible to actually be free to do so.

A real balance needs to be established in defending views to be expressed and not overreacting – Voltaire outlined this – paraphrased as "I disagree with your view but defend your right to express it".

Clear leadership in schools, further education and political direction is needed to once again open up conceptual freedom and rein back on restrictions. Real hate actions are pretty clear – a charter for the "offended" is not needed!!

Michael J Cole

WHERE TO NEXT? – A SOCIAL DRIFT OR FORCED SHIFT

What is the next step in this journey of self-deprivation? With the loss of freedom of thought and expression, the controlling governmental forces (police) over the exaggerated virus, the abasement of Britain in BLM and the ongoing ever creeping and demanding PC and woke campaigns – there will soon be nothing left in terms of democracy and choice.

The latest COVID-19 virus situation has provided Governments (even in the UK where it once had freedom as a by-word) the opportunity to control its people by imposing restrictive laws and enforcing them with police powers. Powers akin to the Stasi regimes which we once challenged. This has been 'justified' by questionable data regarding public health dangers.

Once these powers have been set in motion will they ever be rescinded? Any occasion at the behest of those in power will likely trigger a repeat of this oppression on the basis of "we know what is good for you" or "saving lives"! Not the structure of a free society.

The all prevailing left wing liberal elitist movements have infiltrated just about every organisation. The police, universities, politics (politicians too scared of offending the twitterarti to challenge), public sector, legal institutions, education and parts of the media. These liberals, in name only, demand total obedience and in keeping with the arrogance displayed will not tolerate any objection to their views – any opposite view is regarded as heretical and shut down either by veto or aggressive threats.

This applies to the so called human made climate change – even though it is not supported by science – the debate is settled (it isn't)

A Virus in Society

but nothing is going to stop the rhetoric and the pursuit of policies to change our way of life and standards for the worse and wreck our economies. As the BBC has declared, "it is settled, we will not allow any challenge to our broadcast on any of our channels!"

They are all getting in on the act. Nicola Sturgeon, the nemesis of democracy with her Scottish UDI-camp seems to think she has control over the whole country. By living in Scotland part of the time, I can confirm that Scotland, under the SNP's control for some time now, is a failing waste land. The hospitals are substandard and ill-equipped (personal experience), the police are mismanaged with a confused system of direction, the schools are a shadow of what they once were and the economy is at an all-time low. Politics are corrupt and rule obsessed but find money to support drug taking and public sector spending to the most incompetent levels.

So when I go shopping in Scotland, living on my own, I will be denied food because I won't wear a mask. Who will feed me – what right has any state to refuse me my right to food?!

We don't need these self-serving sanctimonious politicians to tell us how to behave to conform to specialist influence groups. They would get more respect if they really understood social needs rather than latching on to the fashionable empty bedroom social media crowd.

So we now have a momentum of self-loathing in social structures of the whole country. We have weak politicians not able to stand up to this depressing status and actually supporting the crazy policies to control and restrict real freedom and choice and ruin our economy. This in turn will not only retard our progress and standards but will cause disharmony in society. Within twenty to thirty years the damage will be catastrophic – unless the direction of travel changes – and it needs to start now.

I believe the vast majority of people in the UK don't want the culture being forced on us, like George Orwell's 1984, but we are not asked and the method of democracy is being lost to the social media "offended" brigade.

Can someone please stand up for what real democracy is about and not let us be led blindly into the destruction of what Britain used to stand for. Despite its problems and political frailties of the past, it held out for freedom and we held on to a set of decent principles. The tolerance and fairness that is being lost was always here in the majority but unfortunately the mea culpa being displayed now is actually bringing this into question under the oppression of conformity.

This is a time for real leadership in politics, not kowtowing to the minority baying mob culture but standing up for balanced reasonableness. This is truly needed if we are to retain our freedom, enterprise and democratic nature.

3 THE MYTH OF HUMAN MADE CLIMATE CHANGE

Included here are some papers relating to this subject which is referred to in previous sections. They are written from a position of challenge to the rhetoric and "religious doctrine" adopted by supporters of the theory.

These challenges are based upon published scientific, real world data and as expressed in many credible publications.

This subject and its treatment by the institutional establishments tantamounts to a blind following of rhetoric, irrespective of the fundamental inconsistency with the real world data. It has in many minds and protagonists been 'settled' in that there is no argument against it – well there is!

These papers set out to show the disparity and to discredit the basis of the claims and the dangers to economies and, therefore, societies if it is allowed to dominate the agenda.

A summary position based on real data:

'Any slight warming in the current era is not caused by CO_2 or human activities, to try to say it is based on deceit, lies and manipulation.'

Michael J Cole

WHAT GREENHOUSE EFFECT?

This was an experimentally discredited theory in 1900, but now the theory is exploited – why?

Follow the money, trillions of dollars have been obtained (of taxpayers money) by public bodies, instructions and so called research establishments to keep the theory going – even promoting it. Governments have used it to generate tax revenues. The propaganda goes on!

The theory is unsupported by experimental science, observation and historical data. The link between CO_2 and Earth temperatures is not proven – it does not correlate. Nor is the human link of human made global warning (anthropogenic global warming (AGW)).

Real world data does not support it. Moreover, there is no need to change our life styles or economic policies to "save the planet". We, humans, do not have such a galactic scale impact on the Earth or its planetary environment – it balances itself by the solar and physical effects of the universe and the Earth's own systems.

It really is time to promote the real science and stop the distorted propaganda on AGW – and stop the unsupported and unsustainable so-called 'green' energy drive – it isn't 'green' in fact it is just the opposite.

BAND-WAGON OF GREEN CLAIMS

There is a surfeit of adverts all linking their products or services to saving the planet in more and more tenuous and outlandish claims. Totally unjustified in most cases such as vegetarian food products claiming reduced "carbon footprint" but using the same if not more production and distribution costs than meat protein, linking to wind power but plugging into the national grid using all kinds of energy generation bases and ignoring the metal and fossil fuel manufactured infrastructure.

This misleading and frankly disreputable marketing must stop and be regulated. The whole platform of carbon as an evil substance is wrong and nonsense. We are all made of carbon compounds and the atmosphere needs CO_2 to support plant life, food production and life itself.

This whole area needs examination.

The basis of "green energy" is full of misrepresentation and false claims.

This subject is covered in Michael Moore's film, "The Planet of Humans", where it exposes the misleading claims made for various so-called 'green' energy projects – really worth watching. It shows the inadequacy of the power generation, the pollution associated with it (not disclosed by those promoting it), the uneconomic cost (and compared to alternative power sources, including fossil fuels) and most concerning, the financial corruption by those involved and earning money from the exploitation of the concept. This includes the institutions getting 'research' grants, industrial corporations latching onto the money making opportunism, and the individuals, including politicians and celebrities seeking fame and fortune. All this amounts to trillions of dollars of public funds – yes trillions! Consider the following:

1. Wind Power ignores the cost and pollution associated with the infrastructure to build and maintain the windmills. At huge environmental cost it is unreliable, inflexible and unaffordable using massive subsidies. It is not free and not CO_2 free!

2. Biofuels, sounds friendly but really means wood burning. The process at Drax power station in Yorkshire, England is a case in point. It brings wood from South America (cutting down CO_2 absorbing trees) on diesel powered ships halfway around the world to Immingham docks where concrete bunkers have been built along with a dedicated railway to deliver wood chips to Drax. The wood is burned to produce electricity. So, transporting wood around the world under the guise of "biofuel", because it sounds good, replaced the main supply of coal from a few miles away in Yorkshire. It ignores the CO_2 displacement to get the fuel to site and the environmental impact.

   Also, in the above example the wood burning produces more harmful particulates than coal! Additionally, it is uneconomic as it cannot recover the cost of supply and has received over £3 billion in Government subsidies so far!

   And this is called "green"!

3. The next misconception is electric cars – or battery powered vehicles.

   The battery component is a major issue, it requires a hugely polluting process to manufacture. The special metals of lithium and cobalt are difficult to extract by mining. Sites are left in polluted states and in the countries of supply, child labour is used in the process under deplorable conditions. Oil based supplies, on the other hand, are well controlled and by and large managed effectively.

Batteries eventually fail, they reach exponential decay, which means they can no longer hold charge and need to be replaced. This then causes more energy use and pollution. This again is not free and not clean or "green".

The batteries need charging and use the national grid to do so – powered also by fossil fuel generation.

Those driving battery powered cars may feel pious by nor using petrol or diesel but the carbon emissions have simply been produced elsewhere instead of the exhaust pipe!

The economic effect of the policy announced in the UK to not supply petrol or diesel cars by 2030 is full of flaws and the prediction is for catastrophe. Let's examine this:

- No infrastructure is in place or will be by 2030 – no plans on cost provision have been made. Charging points to cover the Country (for over 30 million vehicles at some stage) at a much higher frequency of "filling up" than petrol or diesel, would be in the region of hundreds of millions – for homes and road networks of high density.

    How this would work for apartment blocks and terraced streets is yet to be determined.

    - What about travelling abroad – would there be adequate locations?
    - What about commercial vehicles including 44 tonne trailers – what battery or charging capacity would be

needed and how long would charging take – and at what cost? Not thought through.

- Generation capacity by the grid is already limited and at times short of supply so where will the extra supply come from for vehicle charging? Not planned or agreed!
- The cost of battery vehicles are much higher than petrol and diesel models. Can people afford to transfer and what about second hand vehicles? If a battery needs replacing in 5 or 6 years at a cost of £10,000 or in the case of a Tesla £20-30,000 it is unlikely to compare to current markets and affordability. Not explained!
- Then there is the tax implications. If the Government does not get fuel duty on petrol and diesel and no Road Fund Licence is charged on battery vehicles where does the hundreds of millions of pounds lost come from? When will the public be told? Again, not thought through!

One other consideration is that oil based fuels have become cleaner and development is currently in progress to have emission free oil fuels. Other fuels like hydrogen and mixed power units will be developed which can utilise the fuel station infrastructure already in place. Batteries for travel are inflexible and limited in their application and will be the Betamax of travel – for those who remember video cassette tape recorders – they were an inferior option and eventually became extinct.

Why and is it necessary for us to jump into a policy for battery vehicles now? And without having a detailed, forensically evaluated plan? It won't work, will be impractical and lead to economic ruin and detrimental to our social structure

– particularly disadvantaging those of poorer means. Government should not be interfering with commercial development, this is for business and if there is a marketable product that consumers want then let businesses exploit it. Certainly, the basis of so-called human made climate change is not a reason to embark on such an inappropriate policy with ill-conceived rationale.

In short, the 'green agenda' is not necessary, not green, not economically sustainable and will lead to uncompetitive global positioning. China is happy to encourage the West to pursue these moves as it will pick up the gaps left by these policies as it continues to use fossil fuels (despite what it says at the eco conferences).

The moves to eliminate waste and pollution is sensible and local clean air again is a good and supportable approach. But the claims or justification to stop human made climate change to adopt these unnecessary policies is unsupported by science and will likely have numerous consequences. CO_2 is not a pollutant!

Michael J Cole

CLIMATE CHANGE MADNESS – LET THE TRUTH COME OUT

It is disappointing and now alarming that the man-made climate change propagandists have infiltrated organisations and bodies who should know better. That the blind following of a dogma is accepted and repeated without proper evaluation is, quite frankly, unintelligent.

It is equally concerning that the media in general and some in particular, like the BBC, have closed their minds and coverage to any challenge to the religion no matter how compelling the evidence is to confront the theory.

The science borne out of clear, published and verified data, real world data, demonstrates that the world climate and weather effects and patterns are quite 'normal' within historical parameters. And that the effect of human activities bear no real or impactful damage to the system.

It also confirms CO_2, rather than being detrimental, is actually beneficial to plant and species existence. The recent, since post 2000, slightly higher world temperature is, in fact, good for the Earth and humankind. It is still well below the Middle Age temperatures and is a natural recovery from the cooling period peak of 1450 and 1750.

These patterns are well documented. Also demonstrated is the lack of correlation between CO_2 levels and Earth temperatures. This is a fact and the rhetoric of claiming human-made CO_2 is causing the Earth to warm is disproven.

Current CO_2 level, at around 400 parts per million, is quite 'normal'. The Earth has had levels up to 7000ppm and as low as 160. Below 150ppm plants don't survive as they cannot absorb CO_2 – needed for photosynthesis!

The whole obsession and panic about threats of doom are created by falsehoods and misguided propaganda and flies in the face of real world measured data. It has been compounded by inaccurate and manipulated computer models, so developed to support the human-made climate change theory. It simply isn't true. The IPCC is a culprit here processing propaganda for the theory in spite of the real data – such historical data is available in published articles and included in the suggested reading book list – page 149.

The latest theory of melting polar ice caps and drowning us all defies the laws of physics as the icebergs are already 8/9ths submerged and there will be no significant increase in sea levels.

The other scare stories of acidification and deoxygenation of the seas are just that, scare stories as no scientific evidence supports it.

The real tragedy is that the public as a whole, politicians and the like have swallowed the 'Emperors New Clothes' theory and are changing our economic and social structures to comply with something that humans have no real influence to change. These proposed economic and social adaptations will damage our lives unnecessarily. Who will take responsibility for this when the truth eventually comes out – which it will, let's hope it's not too late.

It is about time this whole topic was opened up and the correct science brought out. The public have a right to know the data and have a balanced and sensible debate.

After all science is about objective and impassionate measurement not emotional and manipulated data to fit a theory.

We need this now before we, as humankind, go over the cliff! As unpopular as this position will be in some closed minded quarters, it is essential.

A MASTERCLASS OF DECEPTION
The Climate Change Con!

Niccolo Machiavelli's proposal that "those wishing to deceive always find those willing to be deceived".

This is especially apt when considering the 'climate change' issue. It was called Global Warming but then it became obvious that it wasn't happening during the 1990's and into the 2000's the title was adapted to be more vague and to avoid the embarrassment of explaining the failure of the predictions.

The deception began in the late 60's and 70's, started with the formation of the Club of Rome. Influential people like Maurice Strong and John Holdren were initiators of 'Green' policies and population control concepts. Rich and well connected but with clear centralisation, socialist and draconian motives they used concepts of environmentalism as a vehicle to pursue their approach and objectives.

Their connections with the United Nations and US Government allowed the basis of the 'Green' agenda to get a foothold. It was later further reinforced by others including, of course, Al Gore, who produced the influential film and concept "The Inconvenient Truth". Although the content has been extensively destroyed in detail by inconvenient facts, it gained traction for the theory.

The origination of the Club of Rome did this for financial reasons and for control, they had no real or proven basis for promoting a global catastrophe and so it goes on today. Followers of this and other such individuals like George Soros, under the guise of philanthropic intentions, continue to push the propaganda – a deception in the face of real world scientific data – to increase their wealth and influence.

This is extended into parties of vested interest, like Universities for grants, commercial organisations, political bodies and individuals, including the likes of Lord Deben, John Gummer, who sits as chair of the Committee on Climate Change setting policies for the UK for which his company gets benefits. It also extends to the IPCC and East Anglia University who have been caught manipulating the data to support the Climate Change theory and continue to get grants for research. The infamous 'hockey stick' graph used to claim a warming pattern was selected to obscure the wider pattern which showed no continued warming. The IPCC and East Anglia University were caught trying to manipulate the reports by emails exposing the fraud.

The whole basis for the deception is to foster an attitude of public fear to keep the momentum for supporting the theory for financial gain and control. The propaganda programme has been one that even surpassed the Nazi campaign.

The media has bought it in the main and the BBC feeds the public with a daily diet of support. No challenge is allowed.

However, the real science remains and the truth will eventually come out!

The deception sounds like a conspiracy theory – in fact it is. Vested interested parties have persuaded many that the theory of human made climate change is real and the science is settled – it isn't. Real world data shows clear evidence that:

1. CO_2 is not correlated to Earth temperatures
2. Human behaviour and industrialisation is not the cause of Earth temperature increases – historical data demonstrates this
3. Current CO_2 levels are less than half of the average over thousands of years and around 5% of peak levels

4. Plants need CO_2 to survive – plant life is more productive at higher levels – it is not harmful to the planet.

The 'save the world' concept is an extension to the deception – a confidence trick about which we should not be fooled. It really is the 'Emperor's New Clothes' concept which should be exposed.

Waste, pollution control and efficiency improvements are to be applauded but to embark upon energy production which has hidden disadvantages on the basis of wrong scientific rationale with ruinous effects is not to be recommended.

To change our, the worlds, economic and social structures on the basis of a deception is madness and a full, proper scientific challenge and review needs to be conducted before real damage is done – not by impending global warming or climate change but by the effect of unnecessary political policies.

Michael J Cole

ABANDON ANY HOPE OF TRUTH ABOUT THE CLIMATE CHANGE TOPIC

The lack of understanding of the real data and the ignorance of those who should do the research and know better is staggering. The media in general and journalists, celebrities and politicians in particular are guilty. It is shameful as they either blindly repeat the rhetoric and the social media stories or carelessly seek publicity for taking "popular" media centric views. In some cases it is done deliberately for financial gain.

In the latter case a classic example is Lord Deben, aka, John Gummer MP. He is Chairman of the Committee on Climate Change, setting policies for Government and yet selling his services to the contractors appointed to engage on such work for these policies to the tune of millions of pounds in fees for his company. This happens and no-one stops it! Conflict of interest restrictions are abandoned for climate change.

The media and especially the BBC, our national broadcaster in the UK, funded by public money, have effectively shut down any challenge to the stance of human made climate change. Indeed, as mentioned in other parts of this book, the BBC has stated in its internal journalistic mandate that no challenge to the theory is to be allowed, "the science is settled" so no broadcast is permitted to present a different position to that of their narrative of human made CO_2 is destroying the planet.

Their so called science correspondent, Roger Harabin, takes every opportunity to repeat and reinforce the Corporation's stance. As a science correspondent he should be obliged to fully evaluate the whole subject, but no, he sticks rigidly to the script. There is wide objection by the scientific community to the alarmist claims of those like Harabin but not a jot of acknowledgement is every made. No alternative view is ever expressed and it effectively becomes a propaganda platform and brainwashing of the public. At one time Lord Lawson would occasionally be allowed on the radio to put arguments against the theory but no longer. Presenters like Johnny Ball and David Bellamy who held opposing views to the theory were ousted from the BBC's programmes because they didn't buy into the BBC's narrative on this subject.

Many celebrity personalities see fit to support that CO_2 is killing us position but it is clear they are not fully briefed on the real world data, again simply repeating the media friendly position.

Politicians too are either ill-informed or lazy in that those following and promoting the rhetoric are not taking a balanced view. They need to have courage to research the facts, all the facts not just those produced by vested interest groups, and present them in a constructive way and not be fearful of challenging the "settled position".

Organisations like the IPCC and others really do have a duty to offer a properly balanced scientifically based view and politicians in a position of influence should heed the same advice. It is incumbent upon such groups and including the media to exercise objectivity to reflect true data and set relevant and sensible policies not abandon this for the sake of distorted analysis, or in today's parlance, fake news.

Michael J Cole

ECO ISSUES – comment

Waste and pollution is one thing but embarking on the madness of a hypocritical energy and commercial revolution is another, with the damaging impact on our life and social structure. The Armageddon fear is totally misplaced.

It is being pursued by ill-informed and excited groups of people with, arguably, time on their hands and who have not thought through the consequences and methodology of implementing what appears to be backward changes to human existence. They don't really understand the impact of their proposed alternative ways of providing energy, travel, doing business or even eating and living.

There is a zealous, religious-like fervour to their enthusiasm. Their hypocrisy is evident and actions inconsistent with their claims – ie driving diesel transit vans to demonstrations – and not flying unless it is to attend a conference to … object to flying!

It really is out of hand and interfering with and interrupting people's everyday lives and businesses is not acceptable. The arrogance and 'we know best' attitude of these people is also astounding. Many attendees are protesting for protesting sake and have no idea about the issues.

The basis for the 'protests' are also misguided. The Earth has existed for billions of years with climate changes at times of extreme dynamics and massive temperature ranges when humans and Range Rovers did not exist. It will continue to do so caused by solar and galactic influences far greater than humans have or ever will cause.

Localised pollution and waste should be controlled and managed but within the context of modern life requirements – we cannot go back to the middle-ages lifestyles – by the way they did burn carbon fuels to keep warm and eat meat!

However, a sensible and proportionate approach needs to be followed. The developing world – still including China, India and soon to be Africa, will not stop developing because certain over indulged members of UK youth and celebrities decide to block Westminster! Government and all MP's need to resist the hysteria and let common-sense prevail.

Christopher Booker, a scientist and journalist, has fully degraded the popularist and distorted policies and action of our Politicians in embarking on a so-called 'green' agenda with drastically negative results (see his many articles) so we must not fall into the trap of responding to such tactics by the 'protest brigade'.

We have a media fuelled situation where a credited scientist with extensive research experience and qualifications like Dr Jennifer Marohasy is ignored in favour of a sixteen year old Swedish girl, Greta Thunberg, with no science background getting exposure for emotional rhetoric to incite children to follow a theory of climate change caused by humans. Surely, a better balance can be found at least on common sense levels and respect for science.

We are not doomed and mad policies to pander to the extremism will not work!

DEBUNKING HUMANKIND MADE GLOBAL WARMING

1. The Earth's temperature is <u>not</u> higher than historical norms – data proves this.
2. CO_2 is not the cause of any temperature change – data proves this.
3. CO_2 levels are normal at ~400ppm and actually good for plant and life propagation and food yields – it is not a poison!
4. Industrial/lifestyle CO_2 generation is not related to Earth temperatures.
5. CO_2 is 0.04% of atmospheric gases and only 4% of 'greenhouse' effect gases – the vast part of 90% is water vapour.
6. The IPCC and other 'agencies' use selective data and misrepresented data together with dubious computer modelling to support their theory – all of which can be shown to be wrong.
7. Media coverage in general and the BBC in particular, has shut down any challenge or alternative projection of the data to show a different picture.
8. The propaganda of the theory has become a 'religion' of acceptance and it is important to rebut this to halt the potential catastrophe of unnecessarily embarking upon changes to our economic and social structure.

All quantifiable by real world, publicised data and evidence.*

*PUBLICATIONS WORTH READING – TO BALANCE THE ARGUMENT
   Gregory Whitestone – Inconvenient facts
   Christopher Booker – Global Warming – a casestudy in groupthink
   Dr Tim Ball – Human Caused Global Warming
   Bjorn Lomborg – The Sceptical Environmentalist
   James Dellingpole – Eco – Fascism (for fun)
   Dr Jennifer Marohasy – Climate Change – The Facts
   Ian Hall – Unsettled Science
   M J Sangster – The Real Inconvenient Truth
   David Bennett Laing – In Praise of Carbon
   Zina Cohen – Greta's Homework

Some suggested books to offer a quick and easy understanding of the real world science to challenge the alarmists rhetoric and to balance the argument – essential reading before a view is taken.

These are by credible scientists who provide serious and researched arguments.

<u>Note</u>

Michael Moore, the noted American film maker, well known left wing, anti Republican and erstwhile environmentalist supporter has produced a film in 2020, "THE PLANET OF HUMANS", exposing the hypocrisy of the 'green' agenda. It highlights the inconsistencies of 'clean' energy (which does not avoid CO_2 production) and this opens up the damaging economic consequences of pursuing such policies.

If one of their own can see this way it's hopefully the start of common sense and the end of the misguided and deceptive alarmist's rhetoric in the name of environmentalism.

ঌঌঌ

Michael J Cole

ESSENTIAL WATCHING FOR ALL WOULD BE ENTHUSIASTIC CLIMATE CHANGE CONVERTS AND 'GREEN' ENERGY SUPPORTERS
SEE THE TRUTH OF THE CON
UNDO THE MYTH.

Michael Schellenberger has come out! He was a committed environmentalist and part of the "green movement" but is now considered a heretic by those of such faith. His view has changed as he better evaluated and understood the impacts and damaging impacts of such suggested policies would have on economics and society.

His book 'Apocalypse Never – Why Environmental Alarmism Hurts us All" is worth a read. It takes apart the alarmist rhetoric and rejects the cataclysmic claims of the theory and dispels the threat to the planet. This is from an "insider" who was part of the movement. "Renewable Energy" is a somewhat false claim and this prompted the Obama Administration to put $150 billion into this green myth.

This is a true revelation. Tom Leonard did an article in the Daily Mail on 23rd July 2020 summarising the book and motives behind it.

CLIMATE IN CONTEXT

The danger is that we, humans, judge climate as weather, that is short term experiences and not longer term trends. Longer term being hundreds, thousands and millions of years. We are not here long enough to experience such long term changes. In such a span, we in our life time, experience good (ie, warm and beneficial) or bad (ie, cold and difficult) life conditions.

As a pattern it needs be considered over the cycles of climate periods which, according to the records take long periods of time. Singer and Avery's book, "Unstoppable Global Warming", puts forward evidence to show there is a 1500 year cycle. This is the event period, + or - 500 years, in which climate pattern encompasses cold and warm periods. Examples can demonstrate this such as the pre-Roman warming period between 750BC and 200BC, the Roman Warming period of 200BC to 600AD when conditions changed from a cold drier climate to a warm and wetter one. Following this came the cold Dark Ages between 450AD to 900AD when life survival was difficult in the cold, food scarce period. After this the Medieval Warming Period (MWP) of 950AD to 1300AD was experienced when food production increased and humans prospered. This was the time that Greenland became fertile and the Vikings colonised it. Then came the Little Ice Age between 1400AD and 1800AD when conditions became difficult with food shortages and disease. Since then the Earth has stabilised somewhat to 21st Century with a slightly warmer trend even though is has shown stops and starts. Other examples are also seen over a large time frame. People living within these times only experienced what they faced and did not see it in context as we can by historical data.

The major Ice Ages also have a regulated pattern. Over the past 400,000 years there has been four deep Ice Ages when temperatures fell to below -30°C and ice sheets spread to North America and to mid Europe. The likelihood is that the Earth will experience a repeat of this in the future.

The driver of this and the climate pattern is the natural processes in play. The sun and solar discharges, the elliptical orbit, tilt and wobble of the Earth's trajectory play significant influence together with the science mechanisms, particularly of cloud formations and the carbon cycle, all of which create the controlling dynamics. And all of which is outside human control. The smaller changes between the large changes are also conditioned by the natural processes – humans are but observers in the galactic theatre.

The main observation here is that cold periods are detrimental to life and to human development and warm periods promote propagation, development and prosperity.

Measurements shows that solar activity correlates closely with Arctic temperatures and has little, if any, correlation with CO_2 concentration levels. A graph is shown below to demonstrate this.

Arctic Temperatures Correlate with Solar Activity Not CO_2

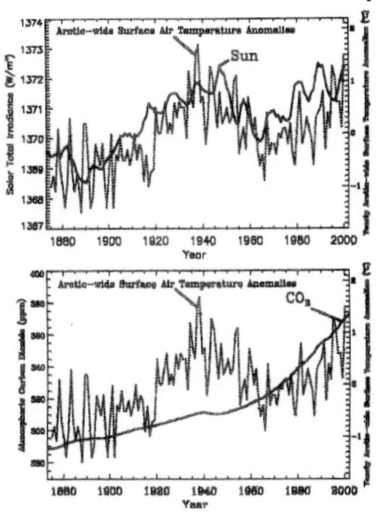

Source: W. Soon, "Variable solar irradiance as a plausible agent for multidecadal variations in the Arctic-wide surface air temperature record of the past 130 years," Geophysical Research Letters 32, 2005.

This is rather compelling to eliminate CO_2 as the source of global warming.

Michael J Cole

A PRINCELY MISCONCEPTION

We now see Prince William embark upon the climate change bandwagon. Who is advising him? Has he researched the real data or is he repeating the rhetoric of propaganda blindly following the 'religion'?

The real data, and there is plenty of it and many publications by credible scientists' who have objectively studied and reported it, shows a different picture – please Prince William and all those who are interested, take time to read this before buying into the distorted doom laden arguments that humans are about to destroy the planet Earth!

The two planks of the theory of "human made climate change" can be dismissed:

1. There is no correlation between carbon dioxide (CO_2) concentration levels and Earth's temperatures.

    Historical and proven data over hundreds of thousands of years have shown that variations of CO_2 and temperatures do not form a linear relationship. Indeed higher levels of CO_2 existed in the past when temperatures were colder than now and vice versa.

    CO_2 at current levels of around 400 parts per million is below the average over thousands of years of 700 ppm and lower than peak levels of 7000 ppm.

    CO_2 is needed for plant life to exist and below 150 ppm absorption cannot take place and plants die! Current CO_2 levels are good for plant propagation and food production – facts not announced by the climate change alarmists!

2. Human activity is not responsible for the Earth's climate or any change. The change is due to factors way beyond human behaviour or life style.

Again, data over long periods of history show this. Before humans existed and before our industrial enterprises existed, the Earth's temperature and natural CO_2 levels varied. There are extensive examples and evidence to show that humans have no impact on this – and so it is today!

No explanation can be made for humans to be responsible for the middle age warming period between 950AD and 1300 AD when temperatures were 2-3°C higher than now and Greenland was a fertile land. No industry existed and Range Rovers were not driven by the Vikings!

Also, no explanation is available linking human behaviour to the cause of the cold period between 1400AD and 1800AD when the Thames was frequently frozen. Again CO_2 levels rose and fell during these periods with no direct correlation to climate temperatures.

The current temperatures and CO_2 levels are within normal/natural levels of fluctuation (as seen by real world date) and the conclusion is that there is no human made disaster looming.

And, the notion of ten years 'to save the planet' is not only illogical but lacks any scientific evidence. Someone on the BBC suggested we had 1 year – this is really becoming irritating madness.

The really worrying issue is the 'accepted' view of "human made climate change". It has become a religion, not supported by real science.

Those pushing this theory, including the Intergovernmental Panel for Climate Change (IPCC), who have taken on the mantra of high priestship have been caught manipulating data and presenting it in a way to confirm the theory. Their computer models are flawed and give exaggerated projections – all of which, so far, have been shown to be wrong – predictions have not been anywhere near actual readings – they are nothing more than deliberate distortions to support the theory.

It is alarming that other bodies buy into the unscientific analysis and ignore the inconvenient facts relating to the subject. The motives for research grants related to this is the real driver of course. Trillions of dollars have already been spent – and more to come, with many corporate enterprises looking to milk it like the snake oil salesmen of yesteryear!

The media has given up on proper critical testing of the theory and the inconsistencies to real world data. The BBC feed us on a diet, repeated daily, of the propaganda and have banned any challenges. Radio and TV programmes are nothing more than adverts for human made climate change guilt.

The real problem is that politicians appear to have fallen for the rhetoric and are adopting policies to change and damage our economies and social structure for an 'Emperor's New Clothes' concept. This is the real catastrophe as it will do more damage than the non-existent effect of human made climate change – however dressed up it is an unnecessary programme which in the future will be seen for what it is – a confidence trick.

A full and proper reappraisal of this misguided topic needs to be conducted. The media has a role to play here – when will the public see the facts to balance their opinions?!

This is an appropriate summary of the topic of human made climate change which highlights the distorted propaganda by alarmists and the unjustifiable momentum generated.

This is the Forward by Professor Lindzen to Christopher Booker's book; Global Warming – A Case Study in Groupthink – published in 2018.

Foreward

By Professor Richard Lindzen

The bizarre issue of climate catastrophism has been around sufficiently long that it has become possible to trace its history in detail, and, indeed several excellent recent books do this, placing the issue in the context of a variety of environmental, economic and political trends, Darwall's *Green Tyranny: Exposing the Totalitarian Roots of the Climate Industrial Complex and Lewin's Searching for the Catastrophe Signal: The Origins of The Intergovernmental Panel on Climate Change* deserve special mention in this connection. Booker's relatively brief monograph asks a rather different but profoundly important question. Namely, how do otherwise intelligent people come to believe such arrant nonsense despite its implausibility, internal contradictions, contradictory data, evident corruption and ludicrous policy implications. Booker convincingly shows the power of 'groupthink' to overpower the rationale faculties that we would hope could play some role. The phenomenon of groupthink helps explain why ordinary working people are less vulnerable to this defect. After all, the group that believers want to belong to is that of the educated elite. This may have played a major role in the election of Donald Trump, which depended greatly on the frustration of the non-elites (or 'deplorables' as Hillary Clinton referred to them) with what they perceived to be the idiocy of their 'betters'.

Booker's emphasis on the situation in the UK is helpful insofar as there is nowhere that the irrationality of the response to this issue has been more evident, but the problem exists throughout the developed world. The situation everywhere has been reinforced by this existence of numerous individuals and groups that have profited mightily from the hysteria (including academia, where funding predicated on supporting alarm has increased by a factor of about 15-20 in the US), but why so many others have gone along, despite the obvious disadvantages of doing so, deserves the attention that Booker provides.

Professor Lindzen was Alfred P Sloan, Professor of Meteorology at the Massachusetts Institute of Technology until his retirement in 2013. He is a member of GWPF's Academic Advisory Council.

4 A SELECTION OF LETTERS

Copies of some of the correspondence sent to organisations, the media and individuals are enclosed.

Some of these evoked responses, such responses have not been included to preserve privacy.

These indicate the willingness of the writer to engage and expound the challenges of manipulation of social change and create the opportunity to consider a different view, rethink and debate.

The intention is to see what the emerging narrative being promoted by the "right thinkers" is in reality and the dangers it brings on society as a whole if the orthodoxy and totalitarianism approach is fully achieved.

Alarmingly, schools are teaching and influencing children at early stages the rhetoric of climate change and human blame and creating worry. This is further exacerbated by the link and propaganda created around Greta Thunberg as a figure of identity.

What is missing is the balance of real science and actual data so there is no real understanding just a 'belief system' and thus providing a distorted view which will affect young people's minds and this is a form of unacceptable brainwashing. We abhorred this in totalitarian regimes in the past.

ઢેઢેઢે

Michael J Cole

Staffordshire England

4 January 2021

Mr Tim Davie
Director-General of the BBC
BBC
Broadcasting House
Portland Place London W1A 1AA

Dear Mr Davie

I thought I would write to you having taken up your post at the BBC hoping you can bring a wider perspective to an important issue.

The BBC's reporting or rather propaganda on the subject of climate change is frankly, bias, unscientific (compared to real data) and distorted. It presents a one-sided picture of human responsibility for any such change which is not consistent with the facts. Many eminent scientists have written books on the subject and presented data not supporting anthropogenic global warming (AGW) but the BBC ignores such information. Roger Harrabin, your science correspondent, is a fully paid up member of AGW supporters and never presents anything that contradicts these views.

The main tenets of AGW can be disputed, namely:

There is no data to support a correlation between carbon dioxide concentration levels and Earth temperatures – in fact historical data demonstrates clearly that there is no connection.

Also, carbon dioxide is an essential part of life on Earth – if levels fall below 150 parts per million vegetation dies – goodbye to food!

Human influence over carbon dioxide production is not the main source of carbon dioxide in the atmosphere and again historical data shows that concentration levels have been much higher when humans were not here and the natural variation in such levels is outside human control.

Indeed, current levels of 0.04% (400 ppm) is relatively low. Levels have been as high as 7000 ppm and average over the past 80m years has been around 900 ppm. There is nothing to worry about the present levels.

This whole issue has been exaggerated and the BBC has been active in promoting this exaggeration. Some time ago the BBC management committee of some kind issued an internal paper stating it would not accept any challenge to the AGW theory and would report no comments to the contrary as "the science was settled". It isn't – at least not supporting AGW.

Lord Lawson used to get air time on the Today programme putting forward a different view but he like others has been banned from your broadcasts.

The Intergovernmental Panel for Climate Change (IPCC) has been a main protagonist of AGW theory but has been caught out distorting the presentation of data along with the University of East Anglia, another proponent of the religion of AGW. The BBC appear to be taken in by such distortions.

Surely, as the BBC is a publicly funded body it should present balanced and accurate reporting. Indeed, for the sake of true science it should offer correct assessments. The blind following of AGW is wrong and it is influencing policies which in some cases will result in catastrophe for our economy and social wellbeing. Vested interests are keen to exploit this distortion of the real position.

I, along with other scientists, believe there is a need to have this topic covered with more balanced care and the BBC has a major part to play in this and dare I say an obligation to do so.

Could I ask you to look at this, research the topic fully (not just listen to the converted) and allow proper and open debate and reporting of this to reflect the true science and help create a more responsible view.

I would be interested in your views.

Yours sincerely

Michael J Cole

A Virus in Society

Staffordshire England

MJC/sr

4 January 2021

Mr Sadiq Khan
Mayor of London
City Hall
The Queen's Walk
London SE1 2AA

Dear Mr Khan

I wrote to the Prime Minister about a transport issue in London and received a reply saying it was the responsibility of the Mayor's office and TfL, so I write to you accordingly.

Enclosed is a copy of my original letter outlining the problem, the congestion caused by road layout changes which did nothing but cause delays and the inefficient use of road space.

I would like to know your comments but primarily would like to see these road restrictions reversed. Can you give me and many other road users confidence that this will be done and a degree of common sense and practicality returned?

In these difficult and pressing economic times we need to rebuild Britain's economy and compete in the aftermath of Brexit and we in business need help and not have hurdles put in our way. Transport, especially in London, needs to flow and this should be made a priority.

I look forward to hearing from you.

Yours sincerely

Michael J Cole

Michael J Cole

Staffordshire England

9 December 2020

Dr L Kendon
Met Office
Fitzroy Road
Exeter
Devon EX1 3PB

Dear Dr Kendon

I heard your interview on Radio 4 Today programme earlier this week about predictions of changing weather patterns and your suggested link to AGW. It prompted me to write to you and pose some questions.

One fundamental question regarding the claim of AGW is why does this theory not coordinate with historical data?

Climate has varied over time together with atmospheric conditions and this has not been influenced by humans or industrial processes so why is this being associated with human activities now? Is it structured by the IPCC which was set up effectively to promote this concept rather than research its possibility?

The blame for climate change and in particular warming is put on CO_2 levels as the main issue but no correlation patterns between CO_2 concentration and surface temperatures has been established – indeed historical data shows the opposite so why is this being identified as a cause?

CO_2 levels at around 400ppm are much lower than the highs of 7000 or averages of around 900 over the past 80 million years – in fact current levels are relatively low. A rise of around 80ppm over the past 150 years is not exceptional (Berner data 2001 shows this and it is cover in Gregory Wrightstone's book, 'Inconvenient Truth').

Also, CO_2 is essential for plant life (below 150ppm plants can't exist) and it promotes food production which, presumably, is good.

David Bennett Laing's book, 'In Praise of Carbon', demonstrates that CO_2 is not a greenhouse gas as its chemistry does not conform but that other gases and particularly water vapour has the largest effect on climate and humans have little influence over water vapour production. You will have seen such data in your own specialist research.

Your predictions use modelling and this raises many questions about the input data and structure and I would make the following comments:

The IPCC models and predictions have always been wrong and exaggerated when compared to actual measured data and it brings into doubt the validity of the input.

There is also an absence of understanding critical mechanisms in the dynamics of influencing events and the IPCC acknowledge this to a degree.

Does your modelling take account of the dynamics of cloud formation/water vapour production and does this follow anything which the IPCC has produced?

Also does your model fully incorporate the carbon cycle and the interaction of CO_2 absorption and release from biospheric, atmospheric, hydrospheric and geospheric systems? Again these are natural systems and are independent of human activities but have huge influences on CO_2 levels and balances.

Your claim of extreme weather patterns and predictions is somewhat at odds with published data.

Heatwaves, droughts, tornadoes and similar events are always well publicised, it sells media, but the measured data does not suggest either the intensity or frequency is much different over time. Gregory Wrightstone covers this in his book and so does Dr Tim Ball in his book, 'Human Caused Global Warming – the biggest deception in history'. Does your data show something different?

The currently slightly warmer levels of around 0.8°C over 150 years is rather small and, of course, this has not been linear in that there was a fall or pause in the 70s and 80s which rather forced the phrase 'human made global warming' changed to 'human made climate change' as the data did not support the theory. So I wonder why we are claiming a disaster for the planet with exaggerated temperature increases when the data does not support this?

Also, a slightly warmer phase would surely be welcomed by many as if this were to happen it is less damaging for life and more pleasant than a cold climate – after all in the UK we all go on holiday to warmer places!

Overall, your predictions could be considered alarmist and it is concerning as it seems at odds with real world and historical data. To make predictions the modelling needs to be complete and integrous so I thought I would write to see if you could satisfy me on that basis.

I hope this is of interest and that you can take this seriously on the basis of one scientist to another.

Yours sincerely
Michael J Cole

A Virus in Society

From: Michael Cole
Sent: 08 December 2020 10:30
To: 'letters@dailymail.co.uk' <letters@dailymail.co.uk>
Subject: ASSAULT AND BATTERY

The Government's policy move to force battery vehicles upon the public as outlined is fraught with disadvantages for the economy and will generate negative impacts on our society and way of life – it is somewhat reckless.

There are numerous hurdles to this including:

Ill preparedness – no infrastructure to support the battery charging requirements – no energy supply capacity development to meet demand – no plan to replace the tax regime applied to oil based fuels.

Unaffordable change over to the higher priced battery cars for the vast majority of people and high costs of replacement batteries when they need changing – and they will.

Inconvenience of recharging requirements and no appropriate solution for commercial vehicles and a huge cost disadvantage to industry.

The high pollution impact of battery production which is a hidden issue but will become a major one in the sourcing of materials and dead battery disposals.

Battery powered vehicles will be the Betamax of vehicle travel as alternatives and better systems will supersede the concept.

We should not force this change. Technological advances for non-polluting oil based fuels will be refined – Petronas has been doing work on this and when successful the present infrastructure of fuel stations will suffice.

It will save on huge costs of battery charging infrastructure and save the pollution of battery manufacture and disposal.

'Clean to drive, dirty to make and replace.' – a fitting slogan for the battery concept.

Markets will drive this change and should do so, not Government interference. The policy for battery vehicles is misdirected, premature and unnecessary.

We, in the UK, will not be able to change over by 2030, nine years from now, and we will regret the waste of money trying to do so. Investment in refined oil products and maybe hydrogen should be supported and this would be a better use of public funds.

China will be happy to see us and the West walk into this (they will not do so despite what they say) and they will mop up our economic failures when these proposed moves fail.

Let's rethink this now before it's too late. The public should think before they buy – those with battery vehicles will realise they have no real second hand value and once the oil/petrol solution is found no one will want one!

Regards,

Michael J Cole

Staffordshire, England

A Virus in Society

XMAS WREATHS REALLY THE RST FIRE RISK TO WORRY ABOUT?

ES living in a association ty in Windsor to remove s from our s they are a

supposed to be by the chief ficer. If we emove them Friday, they taken down. report for all, as there here, there is oke risk with s under and

huge notice boards in the entrance hall are covered in paper notices.

Also there are three or four wooden post boxes in the entrance hall, no doubt stuffed with paperwork.

Yet they want our Christmas wreaths to come down. I am 81 and quite a few are older. We have no one to speak on our behalf.

ROSEMARIE WEBSTER,
Windsor, Berks.

around our front doors. There are chairs with foam seating in hallways and wooden hand rails along halls. Two

aste

BATH about an increase of cost of first-class stamps esting. ities, when my wife and I ting (if you know what that le to her every weekday

rone disease by running an incredible seven marathons in seven days to support his remarkable Rugby League colleague from Leeds Rhinos, Rob Burrow. Kevin is light years ahead of the six BBC award nominees for 2020.

Yes, it has been a tough time for

Are we ready to switch petrol and diesel cars for electric by 2030?

I WILL gladly swap my diesel car for an electric one when I can do the following:
1. Drive from Buckinghamshire to Sunderland fully loaded at night without the need to recharge on the way.
2. Charge the car outside my house. I live in a Victorian terrace house and park my car on-street in a different place every night.
Over to you car

manufacturers and the Government. You have nine years to do it.

GARY BARTLETT,
Chesham, Bucks.

THE Government's policy move to force battery vehicles upon the public as outlined is fraught with disadvantages for the economy and our way of life. It is somewhat reckless. There are numerous hurdles to this including:
1. Ill-preparedness — no infrastructure to support

battery charging; no energy supply capacity development to meet demand; no plan to replace the tax regime applied to oil-based fuels.
2. Unaffordable switch to the higher-priced battery cars for the vast majority of people and high costs of replacement batteries when they need changing — and they will.
3. Inconvenience of recharging and no appropriate solution for commercial vehicles and a huge cost disadvantage to industry.
4. The high pollution impact of battery manufacture, which is a hidden issue but will become a major one in the sourcing

of materials and disposal of dead batteries.
Battery-powered vehicles will be the Betamax of travel as better alternatives supersede the concept. Technological advances for non-oil fuels of any kind are a long way ahead and when successful the infrastructure of fuel stations will suffice.
It will save on huge costs of battery charging infrastructure and save the pollution of battery manufacture and disposal.
'Clean to drive, dirty to make and replace' — a fitting slogan for the battery concept. Let's reconsider

this now before it's too late. The public should think before they buy — those with battery vehicles will realise they have no real second-hand value and once the oil/petrol solution is found no one will want one!

MICHAEL J. COLE, Wolstanton, Staf

HOW on Earth can a renewable energy company claim to be able to charge millions of UK electric cars from solar energy... especially in winter? It is dark for up to 17 hours a day and gloomy for much of the rest! For 40 million vehicles, it just doesn't add up

Mr LYN JENKINS, Cign-yr-glyn, Cardigo

made the cut after 36 holes and only just survived the 54-hole cut, as there was then.

I was walking with the crowd down the 18th fairway, the 72nd hole, with JenKin when I suddenly saw Peter walking the opposite way, away from the clubhouse. He was already changed and looked very smart as he had finished his round much earlier as he was one of the first out. He had a last round of 66 that shot him right up the field.

I shouted to him: 'Not a bad round for an ex-Rockape.'

A 'Rockape' is a member of the RAF Regiment. He shouted

back: 'Were you a Rockape? I replied in the affirmative and he shouted 'Do you drink? Come on then.'

I went with him to the Bollinger tent where we drank champagne and chatted about the RAF Regiment, the training camp at Catterick and the Gunnery School at Watchet in Somerset.

Peter was perfectly natural and I felt I had known him for years.

REGARDING Peter Alliss's supposed attitude towards women, in the mid-Eighties

LEN SLADE, Preston, Lancs.

LETTER OF THE WEEK

WIN our Letter Of The Week Magic Mug, courtesy of Printer Pix. There's one for last our Daily Mail Letter Of The Week design when we hot. To create personalised photo gifts (see example, right), visit printerpix.co.uk, including your full Letter's Editor will announce the Letter Of The Week each Friday. Write to: Daily Mail, Letters, 2 Derry Street, London W8 5TT or email: letters@dailymail.co.uk, postal address and telephone number.

Printerpix

sport per se this year, but surely a national recognition of what Kevin has done — raising £2 million for the charity — is the least one might now demand.

Surely the New Year's Honours List will reflect the admiration we all have for Kevin and Rob, and, of course, Scotland New Year Rugby Union player Doddie Weir, whose name is synonymous with the dreadful MND.

MIKE HOLMES,
Sheringham, Norfolk.

A golfing giant

I FOUND the passing of Peter Alliss a very sad occasion. Apart from admiring his golf commentating abilities, I had many years ago met him through a combination of my cheek and Peter's sense of humour.

It was at the Open Championship, in 1968, won by Tony Jacklin. I was, at the time, a constable in the Lancashire Constabulary and was working at the Open. Peter narrowly

Name supplied, Emworth, Hants

my mum, a keen golfer in her 70s, was recovering from breast cancer. My wife, knowing that Peter Alliss was a favourite of hers, wrote to the BBC to tell him of her plight.

In due course my mum received a handwritten letter signed by Peter Alliss expressing his concern, wishing her speedy recovery and hoping she would soon be back on the golf course.

As a family we have ever since all held Peter Alliss in the highest esteem.

OLLIE PEARMAN,
Kenilworth, Warks

Michael J Cole

From: Michael Cole
Sent: 07 December 2020 12:28
To: 'dtletters@telegraph.co.uk' <dtletters@telegraph.co.uk>
Subject: MORE CLIMATE PROPAGANDA

The BBC is a propaganda centre for the human made climate change theory. It refuses to allow any challenge to this to be exposed on its media, in fact it uses the resources of its publicly funded body to ignore any real scientific data to be broadcast which demonstrates the fallacy of the myth.

It allows doom laden statements to be made and even produces programmes to reinforce the misleading projections.

Today on its Radio 4 Today Programme they had a representative of the Meteorological Office, a Dr Lizzie Kendon, pronouncing that Britain will be devoid of snow within 10 years or so due to human emissions and that we will all be in danger of extreme weather!

This is totally unsupported and the modelling, if any, used to project this must be challenged. Such models used by the Intergovernmental Panel for Climate Change (IPCC) have proven to be wrong on all occasions, are inadequate and are missing vital elements about the natural dynamics of our planet's systems and solar effects. These projections are worthless.

After all, a slightly warmer planet is better for life than a cold one. And CO_2 promotes plant growth and food production!

This type of alarmism is designed to frighten and brainwash the public into believing this nonsense. Variable patterns of short term local weather are quite normal and longer term global climate patterns are also shown to vary by historical data. Such variances happened before humans were here and before industries.

The natural variations are far far greater than humans can influence.

Isn't it about time the BBC and the media in general, balanced the reporting so a true picture can be established. It may also stop us pursuing such unnecessary and damaging policies on the back of climate change rhetoric.

Regards,
Michael J Cole
Staffordshire, England

Michael J Cole

From: Michael Cole
Sent: 03 December 2020 11:02
To: 'letters@dailymail.co.uk' <letters@dailymail.co.uk>
Subject: MORE MISINFORMATION

The BBC was promoting a report claiming an increase in warm weather deaths! This is a distortion of real scientific data which shows more deaths are caused by cold weather. They also claimed the UK was facing 'extreme' weather conditions – simply untrue!

Anecdotal experience is recalled as when growing up in Yorkshire and the Midlands we often has cold spells of more than minus 10°C for weeks and highs of 90°C in summer – we were working the fields of the farm.

This is another attempt to proffer propaganda to brainwash people into promoting human made climate change by misinformation. The BBC presents one side only and ignores the real world data and does not allow opposition to the myth.

Isn't it about time the media exposed this type of unrepresentative reporting and look at the vast amount of data to dispel the theory. There is a huge library of books by eminent scientists demonstrating the misconception – the BBC is shameless in its one sided approach.

Regards,

Michael J Cole

Staffordshire, England

A Virus in Society

Staffordshire England

2 December 2020

Rt Hon Boris Johnson MP
Prime Minister
House of Commons
London
SW1 0AA

Dear Prime Minister

The present proposed Hate Crime Bill should be rejected. It is against freedom of thought, freedom of speech and freedom of expression. It will restrict even the drive for original thought.

This move is an anathema to the principles of British standards of free speech for which we have been known and is against all that, as a nation, we stand for. Also it is against what I thought was the freedom principles shared by the Conservative Party.

This proposed Bill will make criminals out of anyone who is accused of 'offending' someone or even on behalf of another party. It puts the onus on proving innocence rather than guilt. It will give a criminal record to someone even though they will have no conviction but just a record constructed by the police. This really is unacceptable – we fought wars to oppose such oppression and police state authority.

Also, the police would be diverted into responding into any unjustified claim or vendetta instead of focusing on real crime and this would seem unjustified.

Such a Bill should not be crept in under the radar, why not make it public with widespread exposure so the electorate can see the proposal and its implications.

Michael J Cole

I have written to the Law Commission to register my objection and I think you should take action to prevent this from progressing to law and I would urge to help stop this Bill.

Yours sincerely

Michael J Cole

A Virus in Society

From: Michael Cole
Sent: 30 November 2020 14:15
To: 'hate.crime@lawcommission.gov.uk' <hate.crime@lawcommission.gov.uk>
Subject: HATE CRIME BILL

I wish to express my objection to the proposed Hate Crime Bill and as requested I confirm my position.

The outline of the Bill is a major containment of free speech, a cornerstone of the UK's longstanding position on human rights. The drafting is a direct challenge to these rights and by definition can and will make criminals out of a person's views irrespective of any action associated with them.

The definition of "stirring up hatred" is too broad and open to interpretation, such interpretation can be used for any number of malicious reasons by those intent on doing so.

The basis of the Bill puts the complainant in an advantageous position by putting the respondent with the onus of effectively proving innocence. The whole basis of the UK laws was for guilt to be proven. A non-intended 'offence' would be put aside to simply favour the complainant's position that they were 'offended' if this Bill proceeds to law. A, frankly, absurd proposition and open to injustice.

The effect of such a law would require police forces to be engaged in such complaints, frivolous or not, and resources are unlikely to be available. This has been seen already, with a less wide Bill than is being proposed and accounting for around 120,000 such "None-Crime 'Hate' Incidents" where no prosecution occurred. Not only taking police resources from 'real' crime prevention or investigation but causing a huge waste of such resources. It also stigmatises those accused by having a criminal record when no crime has taken place!

There are, arguably, sufficient regulations in place to deal with real 'hate crimes' and more ill-defined and extensive regulations and laws are not needed. This is also against the freedom principles enshrined in British culture and is abhorrent to those who fought against such oppression.

This is a further imposition of public freedom, freedom of thought and expression and is likely to impact negatively on the free press approach which is so important to a free state and this bill should not be pursued.

Please record this and hopefully you will take this into consideration in your assessment.

Regards,

Michael J Cole

Staffordshire, England

A Virus in Society

From: Anglia Holdings
Sent: 01 December 2020 10:26
To: 'letters@dailymail.co.uk' <letters@dailymail.co.uk>
Subject: POLLUTED CONCEPT

All of those sanctimonious and pious drivers of battery powered cars should be mindful of the pollution caused by the battery manufacture and the shameful use of child labour in the mining of its constituent parts.

They are responsible for massive poisonous areas of land in the extraction of chemicals like cobalt and lithium used for battery construction.

Battery cars are not pollution free AND when they no longer recharge (and all batteries fail eventually) there is more pollution caused dealing with the waste – they are not recyclable!

The pollution is greater than oil based fuels.

So, to those pushing this unworkable move to battery vehicles, spare a thought for the damage you are doing elsewhere in the world – to this planet that you claim to care about.

Maybe battery cars should come with a health warning. Check out the facts and then decide if you really believe battery power is the answer!

Regards,

Michael J Cole

Staffordshire, England

Michael J Cole

Staffordshire England

25 November 2020

Direct Communications Unit
No 10 Downing Street
London
SW1A 2AA

Dear Sirs

Thank you for your letter in reply to mine to the Prime Minister. Unfortunately I couldn't read the signature of the Correspondence Officer but I felt I should respond.

Clearly this was a standard copied letter and it was constructed as support for the adopted related policies. However, it did not provide evidence of the need for such policies or explain the impact or practicalities of carrying out these policies.

Firstly, the acceptance, as you state, of climate change as an issue is not supported by real world scientific data in fact the opposite is true, so there is in fact doubt about the need to change our economic and social base. Human activities are not responsible for the vast majority of CO_2 in the atmosphere and CO_2 is not a greenhouse gas as such in fact it is essential for and promotes vegetation. Also CO_2 concentration levels have no correlation to Earth temperatures.

Secondly, the recent policy to move vehicle production solely to battery in nine years time is impractical. Can you explain the infrastructure plans, the battery supply and replacement programmes (including the treatment of the related pollution impact especially in mining the minerals) and the energy generation capacity requirements to meet this timescale? What are the plans for commercial transport?

I am concerned that the 'rose tinted' approach will not work, will impose unnecessary changes and hugely impair our economic prosperity. Whatever you say or hope that this will achieve to boost our economic base is a view not universally shared.

I would appreciate it if you could provide any specific answers to my questions.

Yours faithfully
Michael J Cole

Michael J Cole

Staffordshire England

MJC/sr

16 November 2020

Rt Hon Boris Johnson MP
Prime Minister
House of Commons
London
SW1 0AA

Dear Prime Minister

MISGUIDED POLICY – PLEASE CONSIDER THIS

Your announcements on environmental policies are most inappropriate and this worries me.

Firstly, the concept of human made climate change is NOT supported by science and historical real world data demonstrates the folly of this claim. Despite what he IPCC says (and they have been caught out deliberately distorting data presentation previously) there is no connection between CO_2 and earth temperatures. And CO_2 levels have varied over millions of years at much higher levels than we have now caused by natural events not by human activities.

Secondly, to abolish petrol and diesel vehicles by 2030, 10 years' time, is economic suicide. There is no way that the country could adapt to battery powered transport. Also, the so called 'green' energy is anything but – it has been exposed as inefficient, polluting and unsustainable with regard to usage demand. You really should have the true nature of battery production researched to understand the damage that is caused.

I implore you to rethink this. It will be ruinous for our economy and have a seriously negative impact on society on a number of levels.

Pronouncements like this are misplaced and unless detailed plans are agreed, industries affected have bought in and the public consents it will not work. As it stands it creates great consternation. The claims of an economic 'green miracle' will not happen. This was not in the manifesto and we didn't vote for it.

Moves like this to change a whole industry of vehicle production should be consumer market led not prescribed by governments!

As a chemist I have studied the concept of anthropogenic global warming, AGW, and along with many other leading scientists conclude that the theory is wrong and is being promoted in an alarmist way by those with vested interests and the dramatic policy changes are unnecessary.

Let me urge you to have the subject fully and forensically evaluated and please include those specialist scientists who oppose this approach to get a balanced perspective before we set course for disaster.

The UK will not be competitive in the world markets if we set about this emotionally based direction!

Kind regards,

Yours sincerely

Michael J Cole

PS: Specialist scientists of note relating to climate who have written books worth reading are:

Michael J Sangster
Gregory Wrightstone
Tim Ball
David Bennett Laing

From: Michael Cole
Sent: 10 November 2020 10:34
To: 'letters@mailonsunday.co.uk' <letters@mailonsunday.co.uk>
Cc: Peter Hitchens <clockback@gmail.com>
Subject: UNFOUNDED POLICY DECISIONS

It was reported in the Mail on Sunday on 8th November that Prime Minister Johnson is setting environmental policy as influenced by his girlfriend, Carrie Symonds. I find this somewhat bizarre and contrary to logical assessment and decision making.

Government should not be about fads but rather rational thinking. The real science on climate and authoritative data is being ignored and replaced by rhetoric and misleading, selective use of data to peddle a particular agenda. What Carrie Symonds' qualifications are in climate science is unknown but as a chemist who has investigated the subject a passion for environmentalism is not sufficient to direct Government policy.

I have provided Mr Johnson and Dominic Cummings with suggested reading lists on the subject and sent letters explaining the misdirected propaganda on human made climate change but alas I have receive no response.

Why is this Government and politicians in general so hell-bent on driving the human made climate change bandwagon when the real world data does not support it?

Can we please take an objective and real scientific balance on this subject before we set about damaging our economy and social structure.

Regards,
Michael J Cole
Staffordshire, England

A Virus in Society

From: Michael Cole
Sent: 20 October 2020 13:56
To: 'letters@dailymail.co.uk' <letters@dailymail.co.uk>
Subject: SMART METER NONSENSE CLAIMS

The advert appearing in the press, including The Daily Mail on 19 October, is full of misleading statements. Who is paying for this double page spread by "Smart Energy GB"? Is Government funding involved?

The concept of so-called smart meters (often they don't work or transfer with different energy providers) do not save energy. It relies on people to use less – so turning down your heating will do this – you don't need a meter to tell you that!

Also, the content of the advert is full of alarmist rhetoric. The claim of risking a catastrophe of climate change by implying it is human caused is NOT borne out by real scientific data. Talking of "carbon neutral" is a total misnomer – a catch phrase which means nothing. All activities, even breathing, produces carbon dioxide. It is not harmful and is essential for plant life. CO_2 is not causing a change of climate, it does not cause temperature increases – the real science shows this.

The comment in the ad about the Intergovernmental Panel for Climate Change suggesting the world is going to flood is again an unsupported and alarmist exaggeration. There is no evidence of extreme weather patterns – historical data shows patterns are quite normal, that is, consistent with recorded data over long periods.

It is time these misleading statements are stopped – let them be challenged. It is also time that commercial enterprises stopped linking their products and services to a false premise of human made climate change and making misleading claims.

Maybe the Advertising Standards Agency and the media can be more discerning about what they allow to be published!!

Regards,
Michael J Cole
Staffordshire, England

Michael J Cole

From: Michael Cole
Sent: 21 September 2020 13:52
To: 'letters@mailonsunday.co.uk' <letters@mailonsunday.co.uk>
Subject: INCONSISTENT WITH SCIENCE

The Mail on Sunday included an article about David Attenborough and his recent book and I thought I would write to you as the content, I felt, required a response.

I will, no doubt, upset some people with my statements because I will criticise David Attenborough. He may be a messianic figure to many but his positon on climate and the blame he attaches to human activities for climate change is wrong – it does not coordinate with historical scientific data or with climate science.

The popular view of human made climate change is a premise to treat humans as evil destroyers of planet Earth but is inconsistent with the facts. The propaganda of the theory has taken hold and it is disappointing that figures like David Attenborough and especially him with his interest in the natural world, do not understand the real science involved – either they don't or do and choose to ignore it to make a point. Prince Charles has also been persuaded to expound this theory but again he should heed to the real science, not the artificial rhetoric.

His latest book, "A Life on Our Planet", maybe well-intentioned but on the subject of climate and proposals for economic and social changes will be catastrophic for humanity.

Let's look at the key elements.

The premise that humans are responsible for any climate change is not borne out by measured historical data. Climate has been changing over millions of years between quite wide parameters – ice ages and warm periods and most of this has happened before humans existed and no industry operated. Climate change is, therefore, not connected to human activities. A study of this data confirms this – so why hasn't this been considered by the alarmists? Organic life has come and gone over the period of the Earth's existence and it is quite normal to expect this pattern to continue.

Natural forces create change on a massive level and the biggest influence is the Sun and its radiation emissions. These impact on a whole platform of other complex initiatives which create climate and weather. It is on a galactic scale not associated with humans or any other habitants of planet Earth – so why has this been ignored.

The single cause for climate change is claimed to be human production of Carbon Dioxide, CO_2. Well examining this shows that it does not stack up.

CO_2 currently constitutes 0.04% of atmospheric gases. The concentration in the past has been as high as 0.7% and an average over the last 100,000 years or so has been around 0.08%. All this history precedes humans so the variation is by natural sources and galactic physics.

CO_2 is not a poison. In fact without even relatively small amounts plant life would not thrive. At levels below 0.015% trees and plants would die. Indeed at higher levels the planet becomes greener and greater food production is possible to feed animals and humans!

So, CO_2 is not the devils compound. Nor is it responsible for global warming. The facts show that there is no correlation between CO_2 levels and Earth temperatures.

In short, the science does not bear out the theory of human made climate change and CO_2 is not the evil substance it is made out to be. The phrase 'zero carbon' is nonsense – we are all made of carbon as is all organic matter!

David Attenborough should know better, he should acknowledge the real science. I would be happy to debate this with him or indeed I am sure some other eminent climate scientist would too – many books have been written challenging this theory of human made climate change.

Isn't it also about time the media featured alternative arguments to this theory, especially the BBC who repeat the rhetoric ad nauseam!?

Michael J Cole

The public is being brainwashed by a false proposition and governments are adopting unnecessary and impractical policies to kowtow to alarmists supporting the propaganda, policies which will ruin economies and social structure. It is rather arrogant to think humans have a dominant effect of galactic science (which they don't) and impractical to think they can do anything about the natural changes which are outside their control.

There is an economy built around the climate crisis theory so vested interests are strong but the truth needs to be seen and be available to the public.

Come on David, rethink your stance and consider the real science – others too.

Regards,
Michael J Cole
Staffordshire, England

A Virus in Society

From: Michael Cole
Sent: 24 August 2020 10:40
To: 'letters@dailymail.co.uk' <letters@dailymail.co.uk>
Subject: GREENLAND - THE CLUE IS IN THE NAME!

More exaggerated reports about the state of Greenland. Claims of melting ice have been made by University of Reading representative, Professor Ed Hawkins, all adding to the alarmist rhetoric about human made global warming.

Whilst the ice may well be reducing this should be looked at with the benefit of historical measurement and assessment.

Between about 950 and 1400AD Greenland was fertile, hence its name and was colonised by Vikings who farmed extensive areas. It had little ice and in keeping with the Northern hemisphere at that time it was around 1°C or so higher than preceding temperatures. It was known as the Medieval Warm Period. This was also higher than temperatures after this period as between 1450 and 1750 there was a mini Ice Age when the Thames river froze each year. It is worth noting that the Medieval Warm Period was much kinder to humanity than the following mini Ice Age.

This current state of change is, therefore, quite 'normal' in that climate change and temperature fluctuations have happened constantly in the history of the Earth.

For the record there is no correlation between CO_2 levels or human activities to temperature, climate is caused by natural effects largely influenced by the Sun, solar radiation and natural planetary dynamics. Indeed, when the Vikings lived in Greenland there was no industry and to the best my knowledge they did not drive Range Rovers!

Don't panic.

Regards,
Michael J Cole
Staffordshire, England

Michael J Cole

From: Michael Cole
Sent: 06 August 2020 15:45
To: 'letters@mailonsunday.co.uk' <letters@mailonsunday.co.uk>
Subject: CLIMATE CHANGE TURNING!

There are more and more challenges to the 'human made climate change' theory and the green energy obsession as reality strikes. Eminent scientists have rejected the rhetoric and the alarmism for some time with extensive papers and books available to outline the real science which dispel the claims.

The propaganda put out by the IPCC, Al Gore and vested interested parties are beginning to be seen for what it is – misleading and fear inducing distortions of data and make believe to blame humans for what are natural effects influencing climate.

Now even those who once went along with the apocalyptic scenarios have re-evaluated their assessments. Michael Moore's film "The Planet of Humans" exposes the nonsense behind so-called 'green' energy showing that it is not 'green' and a recent notable convert to real scientific evaluation acknowledging the damage by the green lobby is Michael Shellenberger, who could have been described as an Environmental Warrior. He has now shown that the claims of pending climate doom have been exaggerated and wishes to bring a sense of balance and realism to the topic.

The public really do need to have the real facts put to them and opposing views to the theory. The media and politicians need to be more aware of the scientific community who contradict the 'accepted' position. Politicians have demonstrated that they are in thrall to dodgy science as explained by Dr John Lee in his assessment of COVID-19 and the same is true of their approach to climate change.

The true position need to be fully understood before we pursue policies such as green energy and banning petrol vehicles which will fail and ruin our economy and standards of life!

Regards,
Michael J Cole
Staffordshire, England

A Virus in Society

Staffordshire, England

MJC/sr

16 June 2020

Rt Hon Grant Shapps MP
Secretary of State for Transport
House of Commons
London
SW1 0AA

Dear Mr Shapps

I visited London this week, the first time in six months because of the lockdown attempting to do business and was astounded to experience the road changes on Park Lane.

The route from Hyde Park Corner to Marble Arch had been reduced to one lane from the previous three. A journey, even in rush hour traffic, that would take no more than five minutes took over twenty minutes.

There was a cycle lane which has been introduced, the width of a normal vehicle lane, and now a dedicated bus lane added leaving one single lane for all other vehicles. This caused extensive tail backs and congestion in what was relatively light traffic. When we return to normal times the congestion with be catastrophic.

During this twenty minute excruciating journey we saw one lonely bike rider in this cycle lane and in the single lane left for vehicles rows of delivery vans and other vehicle users were stationery.

What is the sense behind this? It is slowing down travel and whilst idling in the jams wasting time and fuel pollution is also increasing. Who could possibly think this is a good idea? It will deter visitors to London and impair business. Delivery vehicles were queuing up at a time when we need to get the Country back to work.

Also, the route from the West to the City and back, along the embankment, has already suffered from the unnecessary lane restrictions and expansion of cycle lanes – which remain empty for most of the day. Not all of us can or want to cycle and surely this obsession needs to be halted.

I am interested in your view and I would ask you if this construction can be reversed – and quickly?

I look forward to hearing from you.

Yours sincerely

Michael J Cole

A Virus in Society

Staffordshire England

MJC/sr

16 June 2020

Rt Hon Grant Shapps MP
Secretary of State for Transport
House of Commons
London
SW1 0AA

Dear Mr Shapps

ELECTRIC CAR BENEFITS

Your latest announcements about electric cars benefitting by allowances is illogical and frankly nonsense. It is unfair to other road users and tax payers.

Let's look at the folly.

Battery powered cars are not pollution or CO_2 free. You need to consider the whole manufacture, maintenance and repair process to evaluate the impact, not just the emissions when driving.

The production of batteries is enormously polluting, the extraction of minerals, the energy requirements, the destruction and replacement adds up to a high and often higher displacement cost than petrol or diesel vehicles. Whilst the impact of such is originated largely overseas the effect is nevertheless the same on the planet.

And where do you think the battery power to charge such vehicles comes from – yes fossil fuel generation so electric cars are NOT carbon free.

In fact a study by Christopher Booker showed that from build to destruction, including all running dynamics, a Range Rover was half the CO_2 displacement of a Prius!

You really should study the science and not let politics override this or indeed trump common sense.

Why should drivers of electric cars get free road fund licence and be allowed to use bus lanes, enjoy reduced parking charges and pay no congestion charges? They use the same roads and space as other drivers and contribute to congestion and yet, arguably, they pay no contribution to road construction and repairs. Why shouldn't they pay towards this? It is totally unfair.

You really do need to reconsider this biased and unscientific policy.

Also, whilst writing, can I ask why there is this obsession with cycling? Not all of us live a mile from work or round the corner as in London. In fact the explosion of bike lanes in our Capital is constipating traffic flows and impeding overall travel. In the North, you know that place north of Watford, we need cars to do business so why don't you encourage enterprise instead of hobbling it by impairing travel – and boy do we need to rebuild our economy!

And, if these cyclists want to use the roads and enjoy exclusive lanes why aren't they contributing to the construction and maintenance costs (and maybe towards the inconvenience of those of us who are too old or infirm to cycle)? What about cycle licences and what about insurance and also licence plates to identify the riders when they damage your car or jump red lights?

Something for you to consider. I really would like to hear your views.

Yours sincerely

Michael J Cole

PS: I write this not only personally but as Chairman of major consumer goods distribution business that needs its staff to travel to develop trade and carry out goods distribution of millions of miles – a feat we have done very effectively during this COVID-19 hiatus!

A Virus in Society

Staffordshire England

10 June 2020

Editors Carrie Love and Roohi Sehgal
Dorling Kindersley Ltd
80 Strand
London WC2R 0RL

Dear Ms Love and Mr Sehgal

RE: DK FIND OUT – CLIMATE CHANGE

I have read the above book and felt I must write to you. The contents and presentation represents a distorted picture of the subject of climate and indeed is a travesty compared to the real science and data.

You fail to put forward a balanced evaluation and omit to reflect the real world scientific data, which is available, to show the Earth's many cycles of climate and temperatures evident over time. The snapshot analysis in your book presents isolated events which does not give a true context and is clearly designed to engender alarmism.

By considering contextual data over long periods of time it demonstrates that human made climate change is a myth as any effect of human activities have little real influence on climate and past episodes of temperature, weather events and CO_2 levels are not correlated. Such measurements varied highly when human populations and industries did not exist.

You make many assumptions not supported by measurements and omit many important relevant facts on climate patterns and you include misleading statements to support the alarmism which, of course, is inconsistent with the truth. I will cover some of them here to highlight my objections.

You do not mention that during the period between 950AD to 1450AD part of 'Medieval Warming' the Earth's temperature was between 2°C and 3°C higher than at present. Greenland was fertile, hence its name, cultivated by Vikings and to the best of my knowledge Vikings didn't drive Range Rovers!

This position had nothing to do with human activities and as with the normal variation of climate was to do with solar activities and galactic effects of radiation totally outside human influence.

You do not state that CO_2 represents currently 0.04% of atmospheric gases – you infer it is dominant but it isn't – indeed at 400ppm it is less than half of the average over the Tertiary Period and well below the earlier peak of 7000ppm.

This should be explained to put it in perspective. You state that global rising temperatures are as a result of human activity as the largest influence – it isn't and there have been periods in recent times where temperatures have fallen. Take the period between 1350 and 1750 when the Thames froze each winter and fairs were held on its surface. Between 1998 and 2014 temperatures dropped across the planet and did not rise even though industry was in full operation and CO_2 levels had increased slightly. All this demonstrates that there is little relationship between human activity and climate.

You don't explain that water vapour and clouds are the main barrier to surface heating. The science behind this is not understood and it is ignored by the climate change modelling which is really so incomplete as to be useless. You also state that ozone is a blue gas – it isn't, it is colourless, the sky appears blue by light refraction.

In relation to the modelling on which you appear to base much of your views it should be pointed out that this is not an exact science and was created to provide an alarmist view and support for human made responsibilities. The IPCC have presented distorted data in this regard along with East Anglia University who were censured for this. You really should have pointed this out in your book but you didn't.

A Virus in Society

You use Greta Thunberg to help attract attention to your book but she is being exploited to seduce children into believing the rhetoric and she has no credible scientific based knowledge. Unlike many real scientists who have published researched books on the subject you prefer to use Miss Thunberg – you should really have read the real science.

You promote so called green energy but fail to explain the detail and the falseness behind this. To produce wind, solar and water power it requires fossil fuels to manufacture such equipment and this produces high levels of CO_2 and is energy inefficient. In the case of bio fuels, ie wood, it destroys forests which are CO_2 consuming and actually produces more particulates than burning coal. To use one form of energy to convert to another is wasteful and it is better to use the primary form in the first place.

You should watch Michael Moore's film "The Planet of Humans" as it uncovers the fallacy of green energy and the hypocrisy and profit driven motives of those behind it - including Al Gore!

You fail to point out that CO_2 is good for plant propagation and food production. Less people die from warm weather than cold weather. Do you realise that vegetation fails to grow if CO_2 levels fall below 150ppm? There is no correlation of weather patterns and CO_2 levels as you claim and no link between CO_2 and surface temperatures as the data would show if you had taken the trouble to check. Your section at the end of the book entitled "Climate Change and Health" is factually wrong and frankly totally fear inducing nonsense with no scientific support – in fact the data shows the opposite of what you claim!

Interestingly you use a polar bear on the front cover, presumable to imply it is an endangered species dying as the ice floe melt. Well, in fact polar bear numbers are the highest they have been in over 100 years – they are also superb swimmers!

There are many other points on which to criticise your book but in summary it is misleading, disingenuous and contains misinformation. It is clearly written as a 'reference' book directed at children and as such should be accurate and balanced. It is disappointing to see it is presented in such a biased and alarmist way – don't let the facts or the truth get in the way of fiction!

It is clear you have not researched the science or taken note of the real data. Maybe you should consult published material and books such as:

Gregory Whitestone – Inconvenient facts
Dr Tim Ball – Human Caused Global Warming
Bjorn Lomborg – The Sceptical Environmentalist
Dr Jennifer Marohasy – Climate Change – The Facts
Ian Hall – Unsettled Science

It is damaging to let children and even parents be brainwashed by producing such a distorted picture – how about producing another book where you deal with the real science?

Yours sincerely

Michael Cole

A Virus in Society

From: Michael Cole
Sent: 04 August 2020 15:10
To: 'letters@dailymail.co.uk' <letters@dailymail.co.uk>
Subject: BBC (IM)POSSIBLE

The BBC, on Monday 3 August, 'announced' a proposal from an organisation called Possible to ban SUV advertising (and presumably to ban SUV vehicles) to help save the planet.

This was repeated at every 'news' slot.

Who or what is Possible and what authority or credibility do they have for the BBC to feature this pronouncement?

Well, they are a 'human made climate change' supporting body with a web site – a charity listing but not research based! It spends more money than it receives. It's run by 'mission campaigners' – so fits the BBC profile.

As usual the stance of Possible and its rhetoric is all about criticism of the modern lifestyle and trying to destroy it – they have no credible or practical alternatives and the myth of 'green' energy does not stand up to scrutiny – Michael Moore's film "The Planet of Humans" explicitly shows this.

In true BBC style they feature this and get Roger Harrabin, their Environmental correspondent, to comment on it. It puts more fuel on their fire of fear and alarmism about climate. No challenge to the theory is, of course, allowed.

Come on BBC, how about some balanced reporting, your Charter, demands it. Why don't you take note of the thousands of real, qualified scientists who have studied climate who object to the distorted predictions of disaster – why don't you put forward their opposite views that the human activities have little or no effect on climate – it's the sun and natural impacts!

Regards,

Michael J Cole

Staffordshire, England

Michael J Cole

From: Michael Cole
Sent: 21 May 2020 14:06
To: 'justicecommittee@parliament.scot' <justicecommittee@parliament.scot>
Subject: HATE CRIME BILL

I wish to express my objection to the proposed Hate Crime Bill and as requested I confirm my position.

The outline of the Bill is a major containment of free speech, a cornerstone of the UK's longstanding position on human rights. The drafting is a direct challenge to these rights and by definition can and will make criminals out of a person's views irrespective of any action associated with them.

The definition of "stirring up hatred" is too broad and open to interpretation, such interpretation can be used for any number of malicious reasons by those intent on doing so.

The basis of the Bill puts the complainant in an advantageous position by putting the respondent with the onus of effectively proving innocence. The whole basis of the UK laws was for guilt to be proven. A non-intended 'offence' would be put aside to simply favour the complainant's position that they were 'offended' if this Bill proceeds to law. A, frankly, absurd proposition and open to injustice.

The effect of such a law would require police forces to be engaged in such complaints, frivolous or not, and resources are unlikely to be available. This has been seen in England and Wales, with a less wide Bill than is being proposed and accounting for around 120,000 such "None-Crime 'Hate' Incidents" where no prosecution occurred. Not only taking police resources from 'real' crime prevention or investigation but causing a huge waste of such resources. It also stigmatises those accused by having a crime record when no crime has taken place!

There are, arguably, sufficient regulations in place to deal with real 'hate crimes' and more ill-defined and extensive regulations and laws are not needed.

This is a further imposition of public freedom, freedom of thought and expression and is should not be pursued.

Please record this and hopefully you will take this into consideration in your assessment.

Regards,

Michael J Cole

Cambusbarron, Scotland

Michael J Cole

Staffordshire England

11 May 2020

Sent to

His Royal Highness Prince William, Duke of Cambridge
Kensington Palace
London
W8 4PU

Rt Hon Boris Johnson MP
Prime Minister
House of Commons
London
SW1 0AA

Dear Prince William/Prime Minister

THE GREEN ENERGY CON - DISPELLING THE MYTH

Please find attached a copy letter which has been sent to sections of the press which I thought I would forward to you with regard to the serious matter of misconceived so called green energy policy.

I urge you to watch Michael Moore's film as it brings to light the myth of the 'green' claims and the hypocrisy of those promoting it. It is a revelation and this film helps to bring this to light.

I hope you find this informative and interesting.

Yours sincerely

Michael J Cole

A Virus in Society

Staffordshire England

11 May 2020

Also sent to: The Times, Daily Telegraph, Daily Mail, Richard Littlejohn and Peter Hitchens

Dear Sir

THE GREEN ENERGY CON - DISPELLING THE MYTH

A letter the public needs to read!

Please, everyone watch Michael Moore's latest film, "The Planet of Humans", available on YouTube. It is essential watching for any and all would be or committed climate change alarmists.

This film is made by a notorious left wing Democrat leaning USA filmmaker who is (or was), arguably, one of their own 'green' energy supporters.

The film exposes the myth of so called 'green' energy and the hypocrisy of those leading exponents and companies behind the movement.

The claims of "clean energy", "carbon neutrality", "low carbon" and "carbon free" sources are totally dispelled. Wind farms, solar panels and biomass generation is shown to be nothing more than a wide scale confidence trick.

The production of equipment and plants for these schemes use vast amounts of fossil fuels to construct and they all displace more carbon dioxide than if the primary fossil fuel was used for energy generation in the first place!

The process of converting one form of energy to produce another is wasteful in the extreme and totally illogical.

The 'green' energy claims are bogus and in fact the industry, taken in total assessment, is more polluting than the primary industries of coal and gas and is much less efficient and less practical.

Why pursue such policies, you will ask, if they are so bad and the question is answered in the film. It is perpetrated by those with huge financial gains as the prize. Billions of pounds and dollars are being made at the expense of damage to nature and taxes paid by the public. The film exposes this too. It also lays bare the hypocrisy of particular individuals like Al Gore who have hidden their real motives and made vast sums of money on the back of the deception.

The public have been taken in, celebrities eager for 'right on' publicity have been taken in and so too have politicians who have supported such policies to change our energy industries and way of life on a false premise. This approach needs to change.

This film is a turning point to re-evaluate this myth of 'green' energy. The BBC will not show you this as it is so engrained with its bias towards human made climate change and Mr Moore has challenged their belief in a 'green' energy utopia by exposing the real status behind it. It debunks the myth and tells the truth

Please, please I urge everyone to watch this film and then decide if we really should follow this path.

Yours faithfully

Michael J Cole

A Virus in Society

Staffordshire England

6 May 2020

Advertising Standards Authority
Castle House
37-45 Paul Street
London
EC2A 4LS

Dear Sirs

I wrote to you recently about two companies using vague climate change rationale to support their advertisements – they are still doing it so let me try again.

QUORN

They continue to claim their products "reduce the carbon footprint" but I would ask what evidence they have to prove this claim. Published data and assessments compared with other protein products would dispute this. The claim is vague and not supported by any direct comparisons in the advertisement.

EDF ENERGY

My objection to this has been outlined in detail and in summary is clearly misleading and again I wonder what supporting data they have to claim "low carbon" as published data would also dispute this. I believe they use energy supplied by Drax Power Station which is noted for using bio mass which effectively brings wood from South America to burn in Selby – the comparative measurements of CO_2 emissions for the whole process does not support low carbon dioxide production against locally supplied coal.

I wonder why you are still allowing these misleading claims to be made in such advertisements which appear simply to use unsupported claims on the back of climate change alarmism. It is a propaganda which is not supported by scientific measurement.

You have a duty to ensure advertisements are not misleading and that unsubstantiated claims are not made.

Would you be kind enough to let me know what action is being taken?

Yours faithfully

Michael J Cole

A Virus in Society

From: Michael Cole
Sent: 06 May 2020 11:30
To: 'letters@dailymail.co.uk' <letters@dailymail.co.uk>
Subject: CLIMATE CHANGE OPPORTUNISM

We wondered how long it would take before the climate change alarmists jumped on the bandwagon to connect their rhetoric to the coronavirus situation.

And today we hear from the Government Committee on Climate Change who have issued a statement to promote a continued programme to restrict our lives under this green fascism and even suggest that the coronavirus financial help should be directed to so called 'green' businesses to the detriment of others!

What definition of 'green' is is as vague as its data about climate change. They miss the whole point about the balance of our economy and the necessity to have economic success to support our standards of living and the social wellbeing we all want.

This committee is headed by Lord Deben, aka John Gummer MP, who takes fees for his personal company on the back of such advice and policies adopted – how objective is this?

The BBC, in its usual biased approach, delighted in reporting on this and including it in all its news bulletins throughout the day. Roger Harabin, their science correspondent, jumped on this with relish with not a word of challenge which is the normal shutdown of opposition on this topic by our national broadcaster.

We also have the mad and destructive approach by Extinction Rebellion threatening to ruin our economy and way of life by openly promoting actions to damage our Country – shame on them.

It really is about time to limit exposure to such misguided and falsely based approaches and commit to our principles of freedom and business centric focus – at this time to emerge from lockdown it is essential we get all businesses working again – the prospect of a world governed by climate change alarmists is unthinkable.

Regards,

Michael J Cole

Staffordshire, England

A Virus in Society

Staffordshire England

4 May 2020

Peter Hitchens

The Mail on Sunday

Associated Newspapers Ltd

Northcliffe House

2 Derry Street

London W8 5TT

Dear Peter

I hope my letter gets to you. After reading your article in The Mail on Sunday this week I wanted to say your positon on the COVID-19 panic is shared by many.

It is an overreaction in the direction by Government and Lord Sumption's article in the same issue elegantly outlines the situation. The choice is a state controlled society as the will of a minority of parliamentarians and their lackies in the civil service aided and abetted by misguided police forces or a free society based on choice and open acceptance of policies. To do the former is a repeat of the dark days of communism and to mirror China like oppression.

Dan Hodges, in his article, also in The Mail on Sunday, outlines the worrying implications of the monitoring systems being considered to control our individual movements and that of society as a whole. Count me out of this too!

No one has an idea of what the outcome may be or when it will unfold. The economic results of this will be worse than COVID-19 and so will the impact on society's health and well being.

Whilst my company, as one of the prominent health food and vitamins distributors in the UK, has gained recent sales increases the longer term uncertainty and the wider impact across the whole of our economic platform is the real issue and it is difficult to know how to plan for this.

Quality of life is more than just breathing and existing it needs to incorporate all the levels of human interaction and enjoying the benefits of a successful economy.

Please keep up your insightful views.

Yours sincerely

Michael J Cole

A Virus in Society

Staffordshire England

14 April 2020

Advertising Standards Authority
Castle House
37-45 Paul Street
London
EC2A 4LS

Dear Sirs

And yet another advert claiming carbon saving supplies – this time by EDF Energy.

They imply that electricity from them is 'low carbon'. This is misleading. They take a variety of generating sources to supply energy from the grid – they do not have any data to back up their claims or implications.

All electricity generation produces carbon dioxide, the grid takes a mixture. No generator has zero CO_2 production. Nuclear is the lowest after taking into account all aspects of site build, distribution, maintenance etc but none is zero.

These types of adverts, about which I have written to you already, are simply playing on this human made climate change issue – which is disputable in its concept. BUT the issue here is the cavalier approach to the advertising messages. These are not probable and they are aimed at letting the public believe in impacts of their products or services not supported by measurement.

It really is time you sent a message out and stopped these misleading advertising campaigns.

Yours faithfully

Michael J Cole

Michael J Cole

Staffordshire England

14 April 2020

Advertising Standards Authority
Castle House
37-45 Paul Street
London
EC2A 4LS

Dear Sirs

Yet another company advertising their products on the misleading connection to the climate change propaganda.

This time it is Quorn with their recent TV campaign.

They say their products "reduce carbon footprint." This is, presumably, meaning the production of carbon dioxide but the consumers don't get any explanation and nor is there any science to prove this claim.

In fact if you take the production process, transportation, packaging and associated impacts there is no real difference to meat production.

It also ignores the fact that plants thrive and propagate better in high CO_2 concentrations and yields are higher – so their claim is somewhat counter intuitive.

This is simply trying to capitalise on the climate change issue and not truthful or substantiated. I would, therefore, ask you to review this ad and have it stopped.

There is too much obvious opportunism about adverts using this topic indiscriminately to promote products with disregard to the principles of advertising and without real scientific authority. Maybe you should remind agencies of the need to comply with the rules.

Yours faithfully

Michael J Cole

A Virus in Society

Staffordshire England

26 March 2020

Advertising Standards Authority
Castle House
37-45 Paul Street
London
EC2A 4LS

Dear Sirs

OVO ENERGY

I wrote to you recently raising issues about the OVO Energy advertisement and the misleading content and I am sorry to say that it is still being aired on television.

The basis of the advert suggests that they are able to reduce CO_2 emissions by people buying energy from them but, of course, this is unproven scientifically and the critical point is that I believe they take their energy from a variety of generating mediums including, of course, fossil fuels. This undermines their advertisement which is clearly designed to mislead consumers.

Could I ask you to take action on this and I would be interested to know your plans.

Yours faithfully

Michael J Cole

Michael J Cole

Staffordshire England

17 March 2020

Advertising Standards Authority
Castle House
37-45 Paul Street
London
EC2A 4LS

Dear Sirs

OVO ENERGY

I am writing to you with regard to a series of advertisements for Ovo Energy as transmitted and seen on ITV yesterday.

My concern is that they are advertising their energy contracts using climate change as an argument to persuade consumers to take supplies from them. They are using non-scientific arguments and emotional rhetoric to indicate that Ovo Energy will somehow reduce the dangers of climate change. There is no scientific basis for this and this is misleading and irresponsible.

They state that 26% of climate change impact is related to heating and the implication behind this is that this relates to 26% of carbon dioxide which, of course, is totally wrong. The concentration of carbon dioxide is actually 0.04% of atmospheric gases and the UK's contribution to total global effect is something significantly less than 1%. Accordingly, the presentation in this advertisement is inconsistent with world data and is, therefore, designed to offer inaccurate information to consumers.

This type of advertising is wholly inappropriate and I would suggest against the standards set for advertisers. Accordingly, I would request you have these advertisements withdrawn forthwith.

I would welcome notification of your response.

Yours faithfully

Michael J Cole

Staffordshire England

31 March 2020

Peter Hitchens
The Mail on Sunday
Associated Newspapers Ltd
Northcliffe House
2 Derry Street
London W8 5TT

Dear Peter

I entirely agree with your article in the Mail on Sunday this weekend about the exaggerated and arguably sinister reaction to COVID-19.

The worry is that such a controlling police state approach is likely to be triggered in future on the basis of "public safety" or "saving lives" or wait for it, "saving the planet" on the pretext of human made climate change.

This is, as you point out, not what this Country stood for. I no longer feel free in my Country. I did when I grew up through the 50's, 60's and even the politically troubled time in the 70's. My very opinions and expressions can be considered subject to scrutiny to force prosecution for even thinking! The recent police action on this lockdown appears to be another step in this direction.

During the 50's we had several disease scares such as Polio, arguably much worse than COVID-19 and we were simply advised to take precautions and as schoolchildren we were all immunised by doctors visiting the schools to administer inoculations.

Whilst it is sensible to take precautions for this current virus there seems to be a much more damaging concern in that we are invoking Stazi tactics on the public. Some of the public seem to like it (very worrying) and the wrecking of our economy by forcibly shutting down businesses is a whole new level of courting disaster.

The recovery of both these issues is of great concern. Under the present direction of travel the economy will become a model of left-wing lunacy and it will take years to recover, if indeed it will within this generation's life time. Our Mr Corbyn is already claiming victory.

A colleague of mine did his PhD in virus study. He claims the numbers being reported are not accurate or indicative of the seriousness as the infection rates are not really known so the death rate is expressed as a false percentage. Also, not all deaths are as a result or caused by COVID-19. Hysteria is being exhibited.

The lockdown is political – when the crisis is over the Government will claim they prevented a worst outcome – if they didn't do something critics would claim they could have prevented it from being so bad – an easy win, politically!

I understand, from a medical friend of mine, that around 45,000 died of Sepsis last year and every winter Staffordshire loses 4/5,000 people as a result of getting flu. Admittedly vulnerable people are at greater risk but COVID-19 seems less serious does it not?

Professor Sucharit Bhakdi is right, the dangers of the prevention are greater than the virus itself. This hyperbole is another example of a society developing the need for Government to run their lives, to be dependent on central direction instead of personal endeavours. It is also easier for Governments to control people if they are dependent, this has been a cornerstone of left-wing politics and of Marxism.

I am particularly concerned about the effects this will have on business and our economy. I own a large food distribution company and we have enjoyed a short-term gain with increased sales over the past few weeks but I worry about the sustainability of this and indeed the wider implications on other businesses and the consumers at large. Many companies will fail, they will not afford the additional loan repayments that Government is offering as there will be no profit or cash flow to support them. This will eventually affect the whole of our society.

A Virus in Society

You make the point about the media's approach to this and I agree that it is somewhat distorted. The BBC, in particular, is a closed medium, it operates to its own mantra and it favours this sort of large central Government approach to society. Books by David Sedgwick and Robin Aitken adequately demonstrate this and so the BBC is not portraying the issues in an open and honest manner.

This whole episode demonstrates a move to centralising and controlling aspects of our lives and it could be argued that this is a deliberate and even sinister approach. You are not alone in your assessment and views – please keep them coming.

Yours sincerely

Michael J Cole

Michael J Cole

Staffordshire England

13 November 2019

Sir J A Redwood MP
House of Commons
London
SW1 0AA

Dear Sir John

I read your book, 'They Don't Believe You' which I found extremely interesting and I concurred with your stated views and those implied in your comments, particularly about so called manmade climate change and 'thought policing'.

I am not sure I would describe myself as a populist, I am an evaluation based realist and a 'freedomist', if there is such a word. Together they add up to my version of conservatism being small central interference of life, freedom of through and expression and entrepreneurial economic structure with a social conscience.

Thank you for putting a case felt by many, many people I meet both socially and in business, views which all too often do not get reflected by our Parliamentary elite or indeed the media.

I have written many times to the traditional press on such topics covered in your book and had several published letters. As a chemist, one topic which fires my attention is climate change. This is an over exaggerated 'religion' fostered by zealots, maybe in good faith but misguided nevertheless. It really is the Emperor's New Clothes basis and one day it will be exposed, until then I guess I am a heretic.

Bjorn Lomberg's book exposes the statistical manipulations of the data to give the results favoured by the 'warmists'. Christopher Booker's books and many articles deal with the fallacies of the science – which is not settled as the warmists would have us believe. Booker exposed the nonsense about Drax power station conversion to bio mass and the excessive carbon dioxide cost of using it in place of coal from 3 miles away in Yorkshire – and with wood producing more particulate pollution than coal!

This last example of Drax shows the platitude of the argument of those supporters without really understanding the detail or the science or the consequential impact of such a policy. We are in danger of repeating this on so many fronts by moving our policies based largely on emotional rhetoric.

This evidence based approach contradicts the elitist views on many topics and is not made public or at least not enough. We end up getting a doctored version of events as determined by elitist political spokespeople and their supporting media.

The BBC, in particular, is disproportionally biased in its chosen areas. Climate change is one, multi-culturalism is another together with anti-Brexit, anti-Trump and even anti-UK and there are others.

Robin Aitken's book, The Noble Liar, exposes, from his inside experience, the distorted reporting by the BBC in its so called news, political programmes and now even entertainment as they pursue their left wing liberal social agenda. As a public body this is not acceptable.

Let's consider Brexit, I am the majority shareholder and Chairman of a large SME with over £100m turnover and 100's of staff. From a business model we import around 70% of goods from overseas and mainly Europe and Brexit is not a fear. Indeed it will bring us wider opportunities. Project Fear is nonsense – we are only a plague of frogs! Let businesses sort the issues once we leave the EU and its restrictive practices – we will resolve the mechanics of maintaining trade as long as the politicians and the bureaucrats don't muck it up!

The arrogance as you explained of the political elite in 2016 about the Referendum was clear. If only those political figures had visited a pub or even talked to SME leaders like myself rather than the multi-national cohorts who wished to keep us in the EU as its suits their anti-competitive position they would have seen the public's, the electorate's views and maybe understood the objections felt by their dismissive attitude to the leavers.

You are right on all the issues in your book, 'we don't believe you' and it is a recurring theme which will change the face of our politics – it's started now but it will roll on and a real alternative leader is needed – it happened in America and it is needed here.

In many ways I do not recognise a lot about the UK compared to the Country in which I grew up. I don't feel free – surveillance, laws, restrictions and the feeling of repression cause an uneasiness not felt 20 or more years ago. The world order has changed and is changing rapidly with Asia having a huge economic and numerical impact. No doubt this has affected our environment but I do feel we should try to preserve the good fundamental concepts of freedom which was the key aspect of the UK's DNA and for which wars were fought. Hopefully the children of today will recognise this as they mature but it will require political direction and a balanced media in whatever form it comes.

I am not pessimistic, there are many better things helping our life which has occurred over the past 50 years and business teaches good lessons about adapting and investing in the future. What we don't want is to regrade through ill-conceived knee jerk policies. Indeed we can't afford to do this because China will take our lunch.

I hope my thoughts, intended to be supportive, are interesting.

Great book – when is the next one?

Yours sincerely

Michael J Cole

A Virus in Society

From: Michael Cole
Sent: 06 April 2020 11:07
To: 'dtletters@telegraph.co.uk' <dtletters@telegraph.co.uk>
Subject: A TALL TALE!

Two reports caught my eye regarding birds in the UK and providing further hyperbole about so called climate change.

The first was that nightingales would be less frequent in the UK due to ... wait for it ... developing smaller wings so they can't fly from warmer climes to the UK.

The reason for this, so it is claimed, is that evolutionary change has been caused by warmer temperatures over the past 25 years. But evolution change would take much longer – read Darwin no less! Also, between 1985 and 2005 temperatures in Europe actually cooled and the small rise since, less than half a degrees possibly, is not likely to cause such physiological change.

The second was that certain wading birds would not visit as they are staying where it is warmer. It makes you wonder why they would not continue their migration habit if it is warmer in the UK as well!

These are yet more alarmists tales. Not scientific but repeated by the supporters of the man-made climate change brigade. The BBC, in particular, was relishing this by repeating it at every news bulletin throughout the day. Both these stories are at least fanciful so can we please have some common sense.

Regards,

Michael J Cole
Staffordshire, England

Michael J Cole

Staffordshire England

13 March 2020

Ms Zoë Ware
Assistant Private Secretary to H.R.H. The Duke of Cambridge
Kensington Palace
London
W8 4PU

Dear Ms Ware

Thank you for your response to my communication to the Duke of Cambridge.

I felt it appropriate to respond to this as your acceptance of human influence on climate change is worrying. It is worrying because the real scientific data does not support such a premis. The publications I referenced you and the thrust of my submission was to challenge this.

The IPCC position has been shown to be distorted and not representative of the evidential data. Indeed, they were caught out along with the University of East Anglia issuing misleading and selected extracts of data to support their theory which was shown to be a deliberate falsehood. They continue to promote an unbalanced perspective.

The media and the BBC in particular continue to present an unbalanced view on the subject without challenge and yet there are enormous numbers of credible scientists, including myself, who can show that the situation is not a pending doom for humanity and despite the emotional rhetoric is not CO_2 related or caused in any significant way by human activities.

I would urge the Duke and others to read the authoritative literature I recommended in order to gain a realistic and more scientific based assessment to avoid unnecessary overreaction and likely damage to our economy and social structure by adopting inappropriate policies.

Thank you for your consideration.

Yours sincerely

Michael J Cole

Michael J Cole

Staffordshire England

6 January 2020

His Royal Highness Prince William, Duke of Cambridge
Kensington Palace
London
W8 4PU

Dear Prince William

I hope this letter gets to you and you read it as I believe it is an import subject relating to your recent comments and initiatives on climate change.

This 'accepted theory' is not supported by science or real world data – it is not 'settled' as propagandists would have us believed. There are many publications and real scientific data which show a different picture and challenge the human made aspect of it.

As a chemist I have researched the topic and I would urge you to read the relevant books by credible scientists and balance the argument. These are compelling to dispel the alarmists propositions of the theory.

The IPCC and East Anglia University have been caught manipulating the presentation of the data to fit the theory and their modelling is bias and been wrong in every measurement so far compared to real world data measurements. Other organisations have bought into the 'religion' who should know better. Financial motives, including research grants, have influenced this.

The media has shut down alternative views and the BBC feed us with a diet of propaganda instead of offering the public a balanced and objective approach which its charter requires. No wonder minds are persuaded to the human made climate change story.

There is an opposite picture. The science is persuasive – I cannot cover all the points here but enclosed are summary papers of mine. Also, please read the following:

- Inconvenient Facts – Gregory Whitestone/Viscount Monckton of Brenchley
- Human Caused Global Warming – Dr Tim Ball
- Unsettled Science – Ian Hart
- The Skeptical Environmentalist – Bjorn Lomborg

I hope this is helpful.

Yours sincerely

Michael J Cole

Michael J Cole

Staffordshire England

28 February 2020

Rt Hon Boris Johnson MP
Prime Minister
House of Commons
London
SW1 0AA

Dear Prime Minister

This bizarre ruling to make the third runway at Heathrow "illegal" is disastrous. It will open up challenges to any infrastructure project by any crackpot intent on bringing our Country to a halt.

The vote on this runway was passed by an elected Government and Parliament with a 4 to 1 majority so how can this be deemed illegal on the narrowest of considerations.

We can forget HS2, your Scotland/Ireland bridge and any attempt to make this Country prosper. We need extended travel links even more if we are to develop world trade now we are leaving the EU.

We need to get a grip on this and ditch/repeal this carbon neutral legislation to avoid such inhibitions to our economy. I have written to you before about the non-proven scientific basis of so called human made climate change and this whole fantasy proposition will ruin us. Our energy supplies, industry and our economic and social structure will suffer especially in a competitive world if we are to continue with this approach.

Please do something to reign in this madness!

Yours sincerely

Michael J Cole

A Virus in Society

Michael J Cole

From: Michael Cole
Sent: 24 February 2020 14:07
To: 'dtletters@telegraph.co.uk' <dtletters@telegraph.co.uk>
Subject: MORE BBC PROPAGANDA

The BBC2 programme on Sunday evening by Simon Reeve about his trip across Russia was another example of the BBC promoting the decline of planet Earth.

The content was unsupported rhetoric about human made climate change with no real scientific data or balanced evaluation only comment by Reeve and interviewees paddling the emotional saga.

Presumably the public are funding this one-sided approach which is now part of the BBC diet fed to the public at every opportunity.

The Earth has been here for over 6 billion years and yet we are led to believe humans will destroy it in "less than 10 years" as is claimed and as the programme stated "we will all fry".

What utter alarmist nonsense. Why can't the BBC, as a main broadcaster, demonstrate responsibility and offer balanced programme making.

Regards,

Michael J Cole

Staffordshire, England

A Virus in Society

From: Michael Cole
Sent: 19 February 2020 10:34
To: 'dtletters@telegraph.co.uk' <dtletters@telegraph.co.uk>
Subject: UNINFORMED PROTEST?

As a note to all these extinction rebellion mobsters, why don't you do some research on the real science before tearing up people's lawns and disrupting life for those who work hard to create wealth and pay taxes?

If they did their research they would find that carbon dioxide levels do not correlate with Earth's temperatures **AND** CO_2 is actually good for plant propagation and food yields. At the current level of 400 parts per million (that is 0.04% of atmospheric gases) it is well within normal and natural levels – below the average of 700ppm over the past 10,000 years and peaks of up to 7000ppm!!

The popular view of carbon dioxide is totally over exaggerated – it is not a poison!

So, read some actual science books before you embark on terrorism and uneducated behaviour. Presumably those responsible for the criminality at Trinity College will be prosecuted.

Regards,

Michael J Cole

Staffordshire, England

Michael J Cole

From: Michael Cole
Sent: 13 February 2020 15:45
To: 'letters@dailymail.co.uk' <letters@dailymail.co.uk>
Subject: A BBC CHALLENGE

The BBC has announced it is planning a documentary following Greta Thunberg. This will, no doubt, be another opportunity to tub-thump the propaganda of human made climate change.

Will they test her understanding of the science and the real world data or will it be more emotional rhetoric about "saving the planet" or the polar bear?

If the BBC is so convinced of the theory of human guilt and science supporting the focus of carbon dioxide as a problem and correlation of Earth temperatures then why do they not allow challenges to be presented?

How about a documentary featuring scientists who can oppose the theory and present compelling real data to support their position.

Come on BBC, if you have nothing to hide let's have a balanced approach and allow alternative data to be offered to the public. After all your charter demands this!!

Regards,

Michael J Cole

Staffordshire, England

Staffordshire England

5 February 2020

Mr Ross Clark
c/o Daily Mail
Associated Newspapers Ltd
Northcliffe House
2 Derry Street
London W8 5TT

Dear Mr Clark

I read your excellent article in the Daily Mail this week and I thought I would drop you a line on the topic of so called human made climate change and related issues.

Please see a copy of my letter to the Prime Minister and enclosures on the subject.

This whole issue is getting out of hand based on exaggeration and hyperbole. The proposed actions to 'save the planet' are not justified by real world science and the public is being misled.

Also, the media is not challenging the claims enough in order to expose the distortions and misleading presentations of data, the BBC is a culprit in particular. Hopefully you can help rebalance the coverage.

Kind regards,

Yours sincerely

Michael J Cole

Enc: Letter to Boris Johnson, Prime Minister
Paper: Debunking
Climate Change Madness
Environmentalists Agenda

Michael J Cole

Staffordshire England

5 February 2020

Mr John Naish
c/o Daily Mail
Associated Newspapers Ltd
Northcliffe House
2 Derry Street
London W8 5TT

Dear Mr Naish

I read your excellent article in the Daily Mail this week and I thought I would drop you a line on the topic of so called human made climate change and related issues.

Please see a copy of my letter to the Prime Minister and enclosures on the subject.

This whole issue is getting out of hand based on exaggeration and hyperbole. The proposed actions to 'save the planet' are not justified by real world science and the public is being misled.

Also, the media is not challenging the claims enough in order to expose the distortions and misleading presentations of data, the BBC is a culprit in particular. Hopefully you can help rebalance the coverage.

Kind regards,
Yours sincerely

Michael J Cole

Enc: Letter to Boris Johnson, Prime Minister
Paper: Debunking
Climate Change Madness
Environmentalists Agenda

A Virus in Society

Staffordshire England

MJC/sr

4 February 2020
Mr Dominic Cummings
Advisor to the Prime Minister
c/o House of Commons
London
SW1 0AA

Dear Mr Cummings

Please find a copy of my letter to the Prime Minister.

You will see that I have serious concerns about the position on human made climate change, a position of challenge which is supported by many credible scientists and I think it is important we avoid serious changes to our economic structure based on genuinely unscientific proof of man-made influence which will be disastrous for our economy and society.

Hopefully these concerns can be properly evaluated before we embark on policies of such magnitude – a serious and balanced review should be undertaken and maybe you could have some influence in this.

Yours sincerely

Michael J Cole BSc (Hons)

From: Times Books <times.books@harpercollins.co.uk>
Sent: 24 January 2020 10:36
To: Michael Cole: Times Books
Subject: TIMES ATLAS PUBLICATION - MISLEADING COMMENTS

Dear Michael

Many thanks for getting touch with your comments detailed below. We have passed these on to the editor, for consideration in future publications.

Kind regards,

Times Books

A Virus in Society

From: Michael Cole
Sent: 14 January 2020 10:36
To: 'timesatlas@harpercollins.co.uk' <timesatlas@harpercollins.co.uk>
Subject: TIMES ATLAS PUBLICATION - MISLEADING COMMENTS

Dear Sirs

I obtained a copy of The Times Reference Atlas of The World and I wish to make you aware of my disappointment and concerns over the comments on pages 94 and 95 entitled 'Climate II/Climate Change'.

The presentation of your script of human made climate change is totally misleading and promotes the myth of human activities being responsible and temperature correlation with CO_2 levels which is, frankly, wrong. This does not coordinate with real world and historic data and as a chemist I object to this distortion.

You also title one element as 'Threat of Rising Sea Level' and this is not supported by relevant studies and this position is unrepresentative of published data and objective research. Your published graphs are not referenced and again are not consistent with alternative or contextual data.

The position of 'settled science' is not settled and there are many publications by credible scientists which dispute the tenet of human made climate change and the projections of catastrophe.

The publishers should have researched this issue more thoroughly before going to print and hopefully you will revise these comments in any future publications. It is a shame you have spoiled a good and traditional book with such biased comments.

I hope you will find these comments helpful. If you would like reference to publications outlining alternative scientific assessment then please let me know. I would be interested in your views.

Yours faithfully

Michael J Cole

Staffordshire, England

Michael J Cole

Staffordshire England

6 January 2020

Mr D J Trump
President
The White House
1600 Pennsylvania Ave NW,
Washington, DC 20500
United States of America

Dear Mr President

I am writing to you to urge you to continue your resistance to the mad world theory of man-made climate change – hope it gets to you.

The real science does not support it. It is fake news created by zealots of the theory and those making money from it and getting extortionate grants for 'research' – trillions of dollars already wasted.

This is the biggest confidence trick of all time – a real 'Emperor's New Clothes' story!!

A bit of science:

- Carbon dioxide levels are normal at 400 parts per million and is less than half the average over time and it is good for plants! Evidence shows that there is NO CORRELATION between CO_2 and Earth temperatures – the real word data proves this.

- Humankind behaviour and industrialisation have no effective impact on climate. Again data shows that temperatures and CO_2 levels have fluctuated over hundreds of thousands of years when man did not exist.

All this is known, but to support the theory the IPCC and other organisations have manipulated the presentations to distort the picture.

A Virus in Society

The real story is in the real world measurements and it is clear there is no catastrophic event caused by humans! The real tragedy is the unnecessary policies being adopted by Governments to wreck our economies and our way of life.

So, please keep us all sane and don't let the alarmists hoodwink the USA and the world into supporting a false premise.

Enclosed are a couple of papers of mine – the publications referred to are worth reading to give a true perspective of this topic.

Hope this is helpful.

Yours sincerely

Michael J Cole

Michael J Cole

From: Michael Cole
Sent: 02 January 2020 15:25
To: 'letters@mailonsunday.co.uk'
Subject: ATTN: PETER HITCHENS

Attachments: ENVIONMENTALISTS AGENDA.docs; Climate change madness - 161219.docx; DEBUNKING 191219.docx

Peter

Again I find myself in coordination with your views expressed in the Mail on Sunday publication on 22 December so I thought I would drop you a line.

This expressly relates to your accurate and incisive critique of Tony Blair. These things need to be said, you have the courage to do so, shame others don't.

His shameful legacy of damage to this country's economy, social structure and international perception has been and will be long lived and his reputation needs proper scrutiny.

One of my other issues is the misinformation of human made climate change about which I have written to you before and the now religious, over exaggerated mantra which the media appear to repeat endlessly. The IPCC and other organisations with political and financial motives continue to supply misinformation. Politicians have swallowed this and are in danger of inflicting real damage to humans by adopting unnecessary economic and social impositions to our lives.

The real world science gives a different picture but the 'debate' has been shut down as 'settled science' to support the theory. It isn't settled, in fact, and the opposite is true when comparing real world data and objective scientific evaluation.

More exposure of the true position is needed!

The media in general and the BBC in particular are deliberately stopping any challenge to the concept of human made climate change and it is distorting public opinion and governmental bodies. It has been described as the biggest confidence trick in history and I am inclined, from a scientific prospective, to agree with this.

Please find a selection of my papers and letters I have put together and as a chemist I find the publications referred to in my paper 'Environmentalists Agenda – Fake News' compelling in their research and analysis.

Hopefully this captures your interest and maybe you could influence extended articles to put a real world data perspective on the distortions fed to the public to balance the argument.

Regards,

Michael J Cole

Staffordshire, England

Michael J Cole

Sent: 09 December 2019 14:19
To: 'dtletters@telegraph.co.uk' <dtletters@telegraph.co.uk>
Subject: ATTN: THE EDITOR - MAN MADE GLOBAL WARMING MYTH

Dear Sir

The article published on 3 December in the Daily Telegraph by Sherelle Jacobs was encouraging in that at least a challenge to the dogma and religion of climate change has been published.

Attached is my short paper on the subject which highlights the disparity between the rhetoric endlessly repeated by fanatics of the theory and the real world facts. The science is not settled in favour of alarmists, in fact the science demonstrates the opposite.

The worrying issue besides the distorted reporting and misleading doomsday projections is the blind acceptance by organisations and bodies such as the BBC, IPCC (of course), politicians and now the UN. The Secretary General, Antonio Guterres, really overstepped the mark making exaggerated claims which are nothing better than nonsense. These groups should really know better. Either they are deliberately setting out to mislead or they are ill informed and should study the real world data more thoroughly.

A real, objective and honest assessment should be made before we embark on catastrophic economic and social changes which will do significant damage. We in the UK and the West (maybe not the US totally whilst Mr Trump is President) will go lemming like over the cliff whilst the developing countries of China, India et al will not, whatever they say as they will carry on with carbon fuelled technologies. After all they are still competitors of world trade!

The latest announcement from the Madrid Conference claim the Antarctic is disappearing – forgetting it was a forest millions of years ago and Greenland, at the other end of the globe, was fertile and colonised by the Vikings and all this before industrialisation started so clearly not man made!

A Virus in Society

We were told, initially, when the propaganda started that the theory was 'global warming' but this was replaced by 'climate change' when the temperatures refused to rise for almost three decades. Then we had 'acidification of the seas' but this didn't work out either and so now it's 'deoxygenation of the seas' which also has little real scientific empirical support. What's next, a plague of frogs? This is exaggeration upon exaggeration created by the fear lobby.

Rather than worrying about global warming we should be enjoying the current dynamic of the natural temperature status because historical data shows a future ice age will occur as it has ten times over the past 800,000 years. We are 11,000 years into a 15,000 year cycle. This will be more devastating than the ill-conceived projections of global warming and we should abandon the potential ruin of our economies and social structure by following a misguided dogma.

This evidence is clear from data accumulated by various independent sources, including Jouzel, Alley, Lisiecki and Scotese which all show the trending of empirical based data and demonstrates that man-made effects are not proven or possible.

Eventually the 'Emperors New Clothes' theory of man-made global warming will be seen for what it is but let's hope it's not too late to stop the waste of money, which is already in the trillions, or the devastating effect on our societies by pursuing so call anti climate change policies.

For all those wedded to the theory or accepting of it, please read the publications outlining the alternative analysis then reconsider.

Regards,
Michael J Cole

Staffordshire, England

Michael J Cole

From: Michael Cole
Sent: 02 January 2020 14:45
To: 'letters@dailymail.co.uk' <letters@dailymail.co.uk>
Subject: MISGUIDED YOUTH

Greta Thunberg would do better, at 16, to attend school science lessons and read up on the actual data and learn the difference between exaggerated propaganda and real world science.

Those using this poor girl as a mouthpiece for misinformation should stop in order to avoid the emotional platform that this gives the supporters of this theory. It is time the myth of human made climate change was challenged to avoid the disaster of adapting unnecessary and damaging policies to our economic and social structures.

The real data is available so let's stop blindly following the manipulated computer models and misinformation being promoted. The International Panel for Climate Change (IPCC) has been found guilty of distorting the presentation of data to support its theory and governments have the responsibility to ensure they react to proven science and not constructed unscientific modelling.

All climate happenings in recent times are within normal levels despite what the alarmists would have us believe. In this information free world let me urge everyone to read the real science publications and then decide.

It is time the media, in general, and the BBC in particular offered balanced programmes and introduced the real world data instead of the propaganda platform which is being forced upon the public. The real disaster for humankind is not the misleading apocalypse of climate change but the self-inflicted wounds of misguided policies chasing something which does not exist.

The two planks of the theory can be dismissed through real science and historical data namely;

there is no correlation between carbon dioxide levels and earth's temperatures – this can be seen convincingly with actual data showing no linear relationship and

A Virus in Society

human activities have no bearing on earth's temperatures though industrialisation or carbon dioxide production – again this is demonstrated by real world data showing higher levels of CO_2 when humans and industry did not exist.

Incidentally it is worth noting that current CO_2 levels at around 400 parts per million is about half the average over time of 700 parts per million and well below peak levels of 7000 parts per million – so things are quite normal. Below 150 parts per million plant life dies as it needs CO_2 for absorption and at current levels of CO_2 it actually helps plant propagation and better yield of food products. It's a shame the alarmists don't tell anyone about this.

So let's have an open and informed approach from the media to put this whole matter in perspective.

Regards,

Michael J Cole

Staffordshire, England

Michael J Cole

From: Michael Cole **From:** Michael Cole/Sue Rollings [mailto:sue@angliaholdings.com]
Sent: 29 November 2019 14:15
To: 'letters@dailymail.co.uk' <letters@dailymail.co.uk>
Subject: FAO - THE EDITOR - SMART METER ADVERTS

Dear Sir

You published a two page advertisement in the Daily Mail on 28th November promoting so-called smart meters with misleading links and connection to climate change in a clear attempt by advertisers to pressure the public that there is a connection with these meters and the climate change theory.

This connection is unprovable and designed to deceive. A complaint to the ASA has been made.

This surprises me as a regular reader as the Daily Mail has run many editorials and articles against the promotion of these meters yet you publish an advert. Surely the revenue gain is not worth sacrificing your editorial principles!

I am interested in your views.

Regards,
Michael J Cole
Staffordshire, England

A Virus in Society

Staffordshire England

29 November 2019

Advertising Standards Authority
Castle House
37-45 Paul Street
London
EC2A 4LS

Dear Sirs

SMART METER GB

Smart Meter GB is at it again with newspaper adverts promoting the same unconnected link of so-called smart meter with the climate change argument as they did with the ill-conceived TV adverts.

The TV ad was cancelled and I would ask you to review this campaign as well.

It is misleading and makes unprovable assumptions about the meters giving distorted impressions to the public.

Yours faithfully

Michael J Cole

Michael J Cole

Staffordshire England

27 June 2019

Advertising Standards Authority
Mid City Place,
71 High Holborn
London
WC1V 6QT

Dear Sirs

SMART METER ADVERTISEMENTS

I have written to you before complaining about this advert and I have had an acknowledgement that you are dealing with this along with other complaints. However the transmissions are still going out at regular intervals with high OTS and the message is still distorted and misleading in my view.

You will see it starts with young people and butterflies giving the impression that Smart Meters provide an 'ideal life' and it claims a direct link with challenging global warming. The impression and connections are not scientific and any benefit is so remote and variable it has to be ineffective.

Could I urge you to take action on this as soon as possible as this is distorting the public's perception.

Yours faithfully

Michael J Cole

A Virus in Society

Staffordshire England

13 May 2019

Advertising Standards Authority
Mid City Place,
71 High Holborn
London
WC1V 6QT

Dear Sirs

I have seen the TV adverts for Smart Meters promoted by the energy companies indicating that it will combat climate change. They are using the latest euphoria on this subject to persuade people to have them installed. I must object to this.

There is no science to support this. The only possible connection is that by monitoring your properties energy use closely <u>and</u> take steps to reduce your usage then the reduction in energy use could reduce carbon emissions. There are so many variables in this that is not consistent with the claims.

For instance if the energy generation is from wind farms or nuclear then energy usage reduction has little if any carbon benefit. Also, homes using multi-fuel sources not measured by Smart Meters, eg coal/wood fires etc would be unaffected.

This contrived link of Smart Meters and "saving the planet" and "a world free of pollution" as claimed is totally misleading. It appeals to an emotional reaction for consumers with no valid basis.

I would say it is a cynical and opportunistic non-supported claim and inference and I would recommend you have the advert in its present form withdrawn.

Yours faithfully
Michael J Cole

Michael J Cole

From: Michael Cole
Sent: 27 November 2019 10:24
To: 'letters@mailonsunday.co.uk' <letters@mailonsunday.co.uk>
Subject: EVEN MORE BBC PROPAGANDA

Not content with the drip drip daily diet of climate change promotion by the BBC they now send a reporter to the Antarctic to transmit from there.

This allows the reinforcement of the religion of manmade climate change and provide the propaganda of imminent destruction of the world as we know it.

All the predictions so far have not materialised and they won't because the real world science does not support the manipulated data or the doom laden prophecies.

Why is the BBC spending public funds on sending a reporter and the team on this mission? Frankly this self-promotion and challengeable theories are not justified and a better balance of debate on this topic is urgently needed.

Who will stop this zealous following of the unscientific religion?

Regards,
Michael J Cole
Staffordshire, England

A Virus in Society

From: Michael Cole
Sent: 26 November 2019 14:43
To: 'letters@dailymail.co.uk' <letters@dailymail.co.uk>
Subject: More Blatantly Biased Corporation propaganda

Here we go again – the BBC creating an editorial event to push the so called climate change agenda with Greta Thunberg and others editing the Today programme later in December which will, no doubt, further send out unsubstantiated claims of doom.

The BBC is a key protagonist of promoting the project with misrepresented data, exaggerated dogma, misinformed and frankly deliberately distorted biased presentations.

The science is not settled as they would have us all believe. Many credible scientists and publications present the real world data but this is ignored by the religious fanatics of manmade climate change heaping more guilt on people for simply existing. These challenges to the dogma are inconvenient truths but the media in general and the BBC in particular just refuse to let their entrenched position be challenged.

In fact the BBC confirmed their position with an internal meeting and their declared policy reading ... "the weight of evidence no longer justifies equal space being given to opponents of the consensus". This is astounding arrogance and against its charter of balance and impartiality. They didn't want this to be known but it has now been leaked.

Rather than having informed contributors to challenge their so called settled science they prefer to have a sixteen year old schoolgirl and other non-science ill-informed individuals such as Grayson Perry and Lady Hale edit what is supposed to be a factual news programme.

They have banned people from posing opposite arguments and authoritative data that is available to stop them unpicking their distorted presentation of this topic. Johnny Ball, well known presenter of children's science programmes and David Bellamy, a well respected botanist who also presented TV programmes have been banished from the BBC in retaliation of their challenge of the climate change dogma.

Michael J Cole

The real status of this whole subject needs exposing so the mad, blind acceptance can be properly considered in order to stop the brain washing of the public. Solar and galactic influences affect weather and climate, manmade impact is infinitesimal. To change our economic structure for a 'Emperors new clothes' type approach will backfire – it is not practical.

Pollution is one thing but the topic of global warming is something quite different. The data does not support global warming to any serious degree and the projections of catastrophe are nothing more than predications from distorted input data of a modelling system which has been discredited.

So, let's have a balanced and truthful review of this out of control issue. It's not 'climate change' that will cause any catastrophe but the political policies and unnecessary changes to our lifestyle that will be forced upon us and which in the zealots own language will not be sustainable.

Regards,
Michael J Cole
Staffordshire, England

A Virus in Society

OCTOBER 13 • 2019 The Mail on Sunday

Write to: The Letters Editor, The Mail on Sunday, Northcliffe House, 2 Derry Str
Fax: 020 7937 3829 Email: letters

Eco-fanatics should direct anger at China

Instead of Extinction Rebellion disrupting roads and stopping people reaching important hospital appointments, as you reported on last week, why don't they turn their attention to China, India, the US and Germany – countries which emit far more carbon dioxide than Britain. Our emissions are tiny by comparison.

These countries are also high users of polluting coal for their manufacturing industries.

If further restrictions cause Britain to lose even more of our manufacturing base, we could go bankrupt, as we won't be able to compete with other countries. We won't be able to help anybody then.
Name and address supplied

This revolution is being pursued by ill-informed people with, arguably, time on their hands, and who have not thought through the consequences and methodology of implementing what appears to be backward changes to human existence. They don't really understand the impact of their proposed alternative ways of providing energy, travel, doing business or even eating and living.

Their hypocrisy is evident and their actions are inconsistent with their claims – such as driving diesel vans to demonstrations – and not flying unless it is to attend a conference... on objecting to flying! The arrogance of these people is astounding.

It's Charles' right to be King

I was disgusted by the suggestion in one letter last week that it was a pity the nation couldn't decide who should be the next King – Charles or William. Charles will be the next King and he will follow in his mother's excellent footsteps as a dedicated, loyal, hard-working member of the Royal Family. The right to the throne isn't a part of a reality show.
Linda Cragg, Grantham

Kate will make an excellent Queen but I abhor the idea of Charles and Camilla being next in line.
Susan Thorkildsen, Cornhill-on-Tweed, Northumberland

The Earth has existed for millions of years, and there was climate change long before man and Range Rovers existed. It will continue to do so, caused by solar and galactic influences far greater than man has or ever will cause.
Michael Cole, Wolstanton, Staffordshire

There's nothing so dispiriting than people complaining about others trying to change the world for the benefit of everyone. They will be the first to moan when their town floods. As for Douglas Murray, who wrote an article last week criticising the police for letting Extinction Rebellion's protests go ahead, what does he want? For all protest to be quashed? If you fancy that, Douglas, go and live in North Korea.
D. Cleary, London

How many of the current demonstrators requiring

Government action on climate change are themselves paying into or receiving pensions invested by their pension funds into companies that are not compliant with their own ethos? Perhaps they should start there to object and withdraw their support rather than protest in the way they are doing at the moment.
Ian Walton, Bridgwater, Somerset

Could Extinction Rebellion explain how they propose to stop natural climate change as we continue to move out of the latest Ice Age?
Name and address supplied

It's very annoying that they are causing serious civil disruption while posting posed photographs of themselves trying to look like models. They are models, in my mind – models of stupidity.
Mick Ferrie, Mawnan Smith, Cornwall

Michael J Cole

From: Michael Cole
Sent: 10 October 2019 15:17
To: 'dtletters@telegraph.co.uk' <dtletters@telegraph.co.uk>
Subject: ECO ISSUES

Waste and pollution is one thing but embarking on the madness of a hypocritical energy and commercial revolution is another, with the damaging impact on our life and social structure. The Armageddon fear is totally misplaced.

It is being pursued by ill-informed and excited groups of people with, arguably, time on their hands and who have not thought through the consequences and methodology of implementing what appears to be backward changes to human existence. They don't really understand the impact of their proposed alternative ways of providing energy, travel, doing business or even eating and living.

There is a zealous, religious-like fervour to their enthusiasm. Their hypocrisy is evident and actions inconsistent with their claims – ie driving diesel transit vans to demonstrations – and not flying unless it is to attend a conference to … object to flying!

It really is out of hand and interfering with and interrupting people's everyday lives and businesses is not acceptable. The arrogance and 'we know best' attitude of these people is also astounding. Many attendees are protesting for protesting's sake and have no idea about the issues.

The basis for the 'protests' are also misguided. The Earth has existed for billions of years with climate changes at times of extreme dynamics and massive temperature ranges when man and Range Rovers did not exist. It will continue to do so caused by solar and galactic influences far greater than man has or ever will cause.

Localised pollution and waste should be controlled and managed but within the context of modern life requirements – we cannot go back to the middle-ages lifestyles – by the way they did burn carbon fuels to keep warm and eat meat!

However, a sensible and proportionate approach needs to be followed. The developing world – still including China, India and soon to be Africa, will not stop developing because certain over indulged members of UK youth and celebrities decide to block Westminster! Government and all MP's need to resist the hysteria and let common-sense prevail.

Christopher Booker, a scientist and journalist, has fully degraded the popularist and distorted policies and action of our Politicians in embarking on a so-called 'green' agenda with drastically negative results (see his many articles) so we must not fall into the trap of responding to such tactics by the 'protest brigade'.

We are not doomed and mad policies to pander to the extremism will not work!

Regards,
Michael J Cole
Staffordshire, England

Michael J Cole

Staffordshire England

7 September 2017

Mr R Littlejohn
Daily Mail
The Daily Mail
Associated Newspapers Ltd
Northcliffe House
2 Derry St London W8 5TT

Dear Richard

WHAT HAVE WE, THE UK, BECOME AND WHERE ARE WE GOING?

Just had to write to you.

The Daily Mail publication of 6 September 2017 was a focal point of the painful position the UK finds itself in, in so many ways.

Following on from Local Councils 'control' of our lives exemplified by the nonsense about bin collections, surveillance and fines about something as straight forward as refuse collections, which at one time worked perfectly well, todays list shows me we have lost it and the mad men have taken over.

Let's highlight some of these issues:

1. The Archbishop of Canterbury pontificating about how poor Britain is and peddling his left wing views irrespective of the facts which actually show Britain to have a better economy and welfare system than almost every other Country! What about the atrocious conditions in parts of Africa, Middle East and Asia your Reverence?

2. Scotland's First Minister announces even more public funded expense when the Country, considered independent from the UK, is just about bankrupt. I live there part of the time and the health service, education

system and policing is inadequate, incompetent and inefficient. But more money is to be thrown at unaffordable projects.

The housing market has been wrecked by the increases in Stamp Duty and Council Tax banding costs and Ms Sturgeon now threatens increased taxes on middle classes to pay for these mad cap ideas.

Also, petrol and diesel cars will be 'banned' by 2032. What plans are in place to do this in 14 years, what are the alternatives, where is the infrastructure and how much will this cost – and, of course, where will the money come from?

By the way, what happened to the independence argument which based the economy on North Sea oil? Has Ms Sturgeon considered that if the SNP abhors oil, her oil based economy funding won't work?

The lunatics have really taken over the asylum in Scotland!

3. Even getting children to school now risks fines for dropping them off in a car – another way of Councils showing who's boss! Councils don't make proper road layout arrangements for school runs but use their failings to impose penalties on ordinary people trying to live their lives.

Like the waste bins issue the tail wags the dog! It really is about time the public stood up to this – and Government needs to set the tone and control the Councils.

4. The NHS again gets the headlines for failure, incompetence and cover up exemplified by the case of Consultant Peter O'Keefe.

He raised issues of failure in a NHS Trust and patient mistreatment but was suspended under some trumped up false charges of bullying staff so the Trust could hide the allegations. Not only that, they continued to pay him for over 5 years whilst suspended and ended up paying compensation for his dismissal.

Meanwhile this talented Surgeon was not practising his much needed skill, and is still not, whilst the Trust pen pushers carry on! They would prefer to cover up their failings and loose a key staff member at huge public cost than deal with the failings. It beggars belief - how do these people continually get away with it?

5. The Justice System is also in chaos as we discover that someone called 'Nick' made up false allegations about prominent people of sexual crimes and received compensation for being a 'victim' when he was not a victim at all but is allowed to keep the money he received under false pretences and perverting the course of justice! Who says crime doesn't pay?

He should be prosecuted and the money recovered. Those affected by these libellous false allegations should sue for compensation themselves. If 'Nick' can't pay then may be the police forces should as they clearly didn't investigate properly in order for 'Nick' to receive money for his lies.

It is also indicated that anyone claiming to be a 'victim' will get compensation even if proof cannot be offered for any crime.

The Police seem to concentrate on publicity focused cases like this, journalists phone hacking and 'hurt feelings' reports with no holds barred resources but neglect real crime such as burglary. The resetting of priorities is needed or we will see the respect and acceptance of our Police forces fall even further.

We have really lost the plot here. Isn't it about time we dropped this victimisation compensation culture. Again Government and the Department of Justice needs to take a lead.

6. Our education system continues to be in crisis. The Blair Government opened up so call matriculation level access and increased university places and establishments in a numbers driven way to demonstrate that it was supporting education. But it did this without raising or even controlling standards – in fact, arguably, it lowered them to increase student numbers.

By doing so it caused the problem of funding and now student loans reaching, in some cases, £60,000 at an interest rate of over 6% is daunting. This situation needs action.

Yes we need, as a Country, high education standards. We need this to support our economy and to compete with other world powers, especially the likes of China and India. Their education systems are well focussed and other than, say, our top universities we are falling behind. We need to encourage practical abilities in such areas as engineering, electronics/computer sciences and medical research as well as academic subjects. Technical colleges once offered something in practical skill bases but we now have an imbalance which needs resolving.

Resentment of a student population and socio-political unrest will make change difficult so a rethink on education and further education funding needs to be done sooner rather than later.

These are just some issues and dare I say nonsenses which have been covered in print and the news recently. There are many more and especially if you bring in the PC brigade and the 'liberal elite twitteratti' which often causes the Government and MP's to dither about what is right and needs to be done.

We need some clear and decisive leadership from Government and its key Ministers to reflect the majority views of people in the UK. These unaffordable policies and isolationist political dogmas needs side stepping as they only interfere with real progress.

Wealth creation is needed and should be applauded. It pays for social structures through taxes, creates employment (which pay taxes) and improves the wellbeing of all. This should be encouraged and not despised as often it is by the anti-success brigade who hamstring commercial progress, the 'sales prevention team', many whom are in Parliament!

The BBC is biased towards a left wing position and often disparages business, too quick to promote bad news. It feeds people with its anti-Brexit positon, its climate change obsession and dumbing down approach which undermines the UK.

Brexit is a classic case in point. We ask the people, they give their answer and many in Parliament want to impede it. We need to get on with it not find every little issue to interfere with leaving the EU and setting our own course.

The examples covered above indicate the dangers of fixating on minutia and not focussing on the big picture and major issues – we need clear leadership and now!!

Yours sincerely

Michael J Cole

A Virus in Society

From: Anglia Holdings
Sent: 25 July 2019 10:16
To: 'dtletters@telegraph.co.uk' <dtletters@telegraph.co.uk>
Subject: EXAGGERATED WEATHER REPORTS

We have had two days of hot weather and the usual global warmists jump on the opportunity to claim doom and gloom and the end of the world as we know it!

I remember occasional long periods of summer weather in the 50s, 60s and 70s before global warming was invented and it would appear that we seem to have less such periods in recent times.

The BBC again picks on a couple of decent warm days to promote the propaganda of man-made causes and the somewhat irritating exaggerated 'health warnings'.

All this ignores the key influence of solar dynamics and galactic impacts which affects long term climate and short term meteorological affects called weather. The solar issues totally outweigh any man-made elements and we seem to have excluded this from any consideration – maybe it's not fashionable but the science needs to be taken seriously.

It is time we put the whole issue into proportion and stop exaggerating the issue and creating unnecessary policies on such distorted logic.

Regards,
Michael J Cole
Staffordshire, England

Michael J Cole

From: Anglia Holdings
Sent: 25 July 2019 11:12
To: 'dtletters@telegraph.co.uk' <dtletters@telegraph.co.uk>
Subject: A MYTHICAL STORY

A mythical story and a mythical Police Force – how on earth could those officers involved be taken in by such a fantasist as Carl Beech. I thought they were supposed to investigate to establish evidence not take such stories at face value. This is an embarrassment and an utter disgrace. Which officers will be sanctioned?

The Police record continues to fall short and this is another example of poor direction, poor motives and even worse management.

Real crimes go un-investigated and unsolved to the detriment of the public but the Police find resources to prioritise and carry out PC initiatives and 'right on' approaches. This needs to change. Where is the objectivity and where are the senior officers who should be overseeing these work schedules.

The public deserve much better and the additional funding every year on Council Tax bills at several times the rate of inflation for policing precept is clearly wasted.

So focused on PR wins they have lost their real purpose and you only need to look at the misdirection of Cliff Richard and Paul Gambaccini cases which demonstrates the misguided approach which has to be said is at the expense of knife crime, burglaries and disruption to the working public with eco fanatics restricting travel when the Police do nothing!

Mr Johnson has another action item on his list. This is not about numbers it is about how the forces are deployed and the priorities allocated.

Regards,
Michael J Cole
Staffordshire, England

A Virus in Society

Daily Mail, Tuesday, June 18, 2019

email: pboro@dailymail.co.uk

LETTERS

y pyjamas, I doodlebug

street were just shells. Over the next few days, I went with my father back to the remains of our house to rescue what possessions we could find.

My father found us another house a few miles away, and we pushed our belongings there using a borrowed costermonger's barrow.

For a few months, my mother, baby brother and I went to South Wales to stay with family while my father continued working at a London hospital. I began having nightmares and developed a stammer.

In 2016, I returned to my childhood street with my wife and daughter. We called at Hackney records office and they provided us with pictures of the explosion of June 18, 1944. I found out for the first time that 60 people had been killed. I realised just how lucky my family had been.

This month, as we remember the D-Day landings and the sacrifices made by our Armed Forces, I have a special memory of my own.

It is my 86th birthday on June 21, and I have made the most of the 75 years that have followed since that tragic day.

Robert J.S. Long, Tiptree, Essex.

Unwatched and unwanted

SO 21 shows on BBC Scotland have recorded zero viewers — absolutely no one.

C'mon, Nicola — the channel was created because of the political pressure you and the rest of the SNP establishment piled onto the BBC.

Most of us didn't want it, nor its £30million annual costs. It would only be polite for you to watch occasionally.

MARTIN REDFERN, Edinburgh.

Blame the parents

NO child of today faces real poverty (Letters)? Rather than a lack of material wealth, youngsters are facing emotional poverty.

Parents are too busy with their own needs to give time to their children. Many working parents are too tired and overload their children with gadgets so they do not have to amuse them.

Others misuse the family cash on drink, drugs, beautifying themselves and going out.

I am not suggesting we should return to hard times, but we need to learn to value what we have and teach our children to do so.

Parenting is hard work and if children lack boundaries no wonder they suffer emotional and mental health problems.

There are children living in poverty, but much of this is due to cash mismanagement by their parents.

PAT KEATING, Newport, South Wales.

Spoiling for a fight?

AMERICA has accused Iran of attacking two oil tankers in the Gulf of Oman.

These tankers were Japanese and Norwegian. What on earth would be Iran's motive for carrying out such an unprovoked attack?

This has all the hallmarks of the Gulf of Tonkin incident which was fabricated to start the Vietnam War.

PETER CALDWELL, Uphall West, Lothian.

Genuine poverty

IT'S a travesty that some people do not recognise there is genuine poverty in this country.

After the war, no one had anything. Nowadays, like it or not, computers and smartphones are a necessity for children.

When some parents fail to manage their income, it's not the children's fault, but they are the ones who suffer.

I experienced hard times when my parents, who were smokers, did not have enough money to pay the electric bill and there was no hot water. We all should make sure that children don't suffer in this way and perhaps be a little bit kinder to people who aren't good at managing.

LINDA BURRIDGE, address supplied.

Carbon calamity

THE Government commitment to radical zero carbon living by 2050 is another example of giving in to zealous, ill-informed youth and celebrity protesters. We would lose our world competitiveness by following such a policy.

Mrs May's legacy will not be one to celebrate. Like it or not, we need to compete. Other countries will not be held back by our soft approach to wealth creation.

We live in a consumer world, and for us to let others produce goods for us and claim we are not engaging in nasty processes ourselves is false.

Drax power station in Yorkshire brings in wood pellets all the way from South America. The coal it used to use was mined a few miles away from the site.

Yes, we need to control waste and pollution, but we should not overreact and change our way of life at such a fast pace.

Could you imagine if we had agreed in 1990 that by 2020 we would have no carbon fuel vehicles?

MICHAEL J. COLE, Wolstanton, Staffs.

Going for growth

AT last! A constructive project straddling the Anglo-Scottish Border.

The Borderlands growth deal will see councils in southern Scotland and northern England collaborate on transport and other projects — and is set to include Northern Ireland as well.

These council leaders can see that the existing Border between

CELEBRATE LIFE OF A LOVED ONE

HAVE you lost a relative or friend in recent months whose life you'd like to celebrate? Our column on Friday's letters page tells the stories of ordinary people who lived extraordinary lives. Email a 350-word tribute to: lives@dailymail.co.uk or write to: Extraordinary Lives, Scottish Daily Mail, 20 Waterloo Street, Glasgow G2 6DB. Please include a contact phone number.

Today's poem

DEAR MRS GRAY

Dear Mrs Gray
From out in the sticks,
What a wonderful day —
You're seventy-six!

And although you have
told us
You're poorly of late
And you don't make a fuss
Or claim help from
the state,

And, of course, we regret
You have been very ill,
You are still in debt
As you've not paid
your bill.

Yes, we're sorry to hear
That you are not eating
And that you're in fear
Of having no heating,

And the only solace
Is your television
And you say you
can't face
Going to prison.

And we know you

Our shining star who never gives up

UP TO the age of five, our granddaughter Rachael was healthy and bubbly. Then she developed epilepsy, which over the years has become worse. Two years ago, when

syndrome and at the age of 24 has started work in a cafe (Mail). For 12 years, Rachael has worked unpaid at a charity shop in Derby and

Michael J Cole

From: Anglia Holdings
Sent: 25 July 2019 10:18
To: 'letters@dailymail.co.uk' <letters@dailymail.co.uk>
Subject: WHO'S IN CONTROL?

I have written before about the uselessness of our Police Force, the misguided approach and misaligned policies and clearly it goes on.

They have given up on drugs, on chasing burglars, knife crime and now policing the streets.

Why do we let a few disrupters commandeer our streets and why do the Police let them get away with it? These protestors claim it's about climate change (and it's difficult to know what their protests are about) but this is just an excuse to hide the real motives of anti-capitalism.

Most of these making up the numbers are youths, appear to be school pupils or students, on holiday now so they have nothing else to do. They are orchestrated by core leftist, anti-business protagonists. They disrupt lives of those of us who are in work paying taxes to keep these protestors able to protest and providing the freedom for them to do so.

They have the right to demonstrate but not interfere with other people's everyday business and to impair essential travel. Why do the Police not take action to keep the roads open? Even the Prime Minister was held up in London yesterday and surely measures can be taken to stop this unnecessary and ill-conceived interruption of decent people's livelihoods – have the Police given up on this as well or have they bought into the disruptive principles on the basis of political correctness?

The BBC give these factions airtime and excuses are made by some of their reporters all on the platform of global warming and they seem to celebrate the celebrities who attach themselves to these causes. The hypocrisy here is astounding by behaving differently to what they demand other people to do.

A Virus in Society

As a country we and our economy are being held back by misguided PC driven politicians and public sector bodies in many areas. This needs to change and hopefully our new Prime Minister will kick start the process.

Regards,
Michael J Cole
Staffordshire, England

Michael J Cole

From: Michael Cole
Sent: 12 June 2019 13:21
To: 'letters@dailymail.co.uk' <letters@dailymail.co.uk>
Subject: A DISASTER MOVIE

This latest move by Government and Theresa May to commit the UK to radical zero carbon living by 2050 is another example of giving in to zealous, ill informed, largely youth and celebrity protagonists. It ignores our economic requirements and world competitiveness that we will surely lose by following such a policy. Mrs May's legacy will not be one to celebrate.

We are in danger of living a disaster movie script in true Hollywood style, conditioned by the religion of climate change and degrading those with opposite views as heretics. The result of such a position will change and ruin our standards of living in the UK. The economic hit will happen if we follow this and it will require a disentanglement of the infrastructure and regulations for us to get back in the global race and our society will lose out accordingly. Like it or not we do need to compete.

China, India and the Far East are already challenging in economic terms, Africa is next and these development areas will not be held back by our 'soft' approaches to wealth creation. We live in a consumer world and the production of goods are essential and for us to simply let others continue to produce goods for us and claim we are not engaging in 'nasty processes' is false – we do this now so how much more will we need to do this as we move to the 2050 cliff edge?!

The science behind any change in climate is not conclusive, still! The projections are just that, models generated by input data. Lomberg's book covers much of these inaccuracies. The propaganda surrounding the arguments are by and large exaggerated and emotional - polar bears are not dying out! The Armageddon scenario of overheating planet is false.

Frankly, silly projects get undertaken. Look at bio mass energy. An example here is Drax Power Station in Yorkshire which brings in wood pellets from

South America – felling trees which capture carbon – shipping them round the world at great energy use to the UK where extensive infrastructure has been built and to burn in the furnaces. They don't tell you that wood discharges more particulates than coal but still call this 'green energy'. The coal that it used to use was mined a few miles away from the site and the Government subsidises this because it is loss making. We are told it's bio fuel so it's good but the truth is different. Chris Brooker has produced an excellent article on this in the Daily Mail.

Much of climate and resulting weather are largely functions of solar and galactic events and impacts – which is beyond human control. The earth has had vast temperature changes over millions and billions of years, before Range Rovers existed and it will endure such changes in the future. The variations we have seen within the last 100 years or so are miniscule in context and as a chemist I am convinced the status is not life threatening today or even likely in generations to come. We are totally overreacting to something that has become emotional and a cause celeb for many.

Empirical evidence is being ignored and manipulated and our Government and other bodies are unduly allowing themselves to be influenced, maybe for the sake of popularism or even tax generation opportunities.

Yes, we need to control waste and pollution but not overreact and change our way of life at such a pace – 2050 is only 30 years away and could you imagine agreeing in 1990 that by 2020 we would have no carbon fuel vehicles?

This policy will not work economically or socially by taking such a revolution of change of all aspects of our lives. Already car plants and employment are seeing the first signs of decline and other industries will follow and other world economies will pick up these at our expense. We cannot go back to a 'middle age' life and in the end our people will not support it as they are adversely affected.

The active elements of the fear programmes pushing for such policies should be balanced by the needs of the majority and economic and social sense needs to take priority.

Let's not live in the disaster movie world, let's have a practical and pragmatic approach.

Michael J Cole
Staffordshire, England

A Virus in Society

From: Michael Cole
Sent: 20 May 2019 11:55
To: 'letters@mailonsunday.co.uk' <letters@mailonsunday.co.uk>
Subject: ATTN: PETER HITCHENS - SUNDAY'S ARTICLE

Dear Mr Hitchens

You have again hit the mark with your article in the Mail on Sunday this weekend. The perilous state of our Parliament and governance is clear – clear to all except, apparently, those who serve in it!

Also, the points you make about Syria and WMDs in Iraq, the Irish situation and the likely cover up of the truth expose our Government's failings and do little to promote confidence in the strategies or the integrity of our politicians.

You have consistently pointed out the flaws in the way our social structure is going and arguably deteriorating and highlighted the truths that our elitist politicians don't want to hear. Please carry on, don't let them shut you down as you speak for the silent majority in our Country and it is essential we have a voice.

Please keep going!

Regards,
Michael J Cole
Staffordshire, England

Michael J Cole

Daily Mail, Monday, May 20, 2019

LETTERS — Write to: Daily Mail Letters, 2 Derry Street, London W8 5TT
email: letters@dailymail.co.uk

Don't fall for the climate fears of the eco zealots

ECO zealots, aided and abetted by some elements of the Government, will ruin us all.

Our cars will be worthless when petrol and diesel engines are banned, and our houses will have no resale value if they don't comply with eco regulations.

We will not be able to eat what we like and our whole lives will be subject to control. Companies and investments will be worth nothing unless they comply with rules set by fringe enthusiasts.

This may sound like something out of a Big Brother film, but we are already on this path. Gullible MPs are falling for unsound projections of doom about global warming. But it will lead to a different doom if we follow such an extreme approach.

The developing world with its industrial economies will not slow down as they build wealth for their citizens. When Britain is left behind, who will support our eco salts then?

Who will be the 'heretic' in Government to bring some sense and balance to this issue?

MICHAEL J. COLE,
Wolstanton, Staffs.

Ruined by tourists

IT'S welcome news that environmentalists are trying to halt airport expansion in the UK, but air travel is only one part of the worldwide expansion of tourism that is causing great harm to our planet.

At any international airport, you will see large groups of nationalities on the move around the world who only a few years back would not have been able to afford travel outside their own country.

Who are we in the West to tell these new adventurers not to travel in order to save the planet?

For more than 50 years I've been lucky to travel every year to Sri Lanka, one of the most beautiful islands in the world.

Travelling around used to be on rickshaw or bicycle on dirt roads, but now it's all cars and traffic jams. This once beautiful tropical island is now not much different from Malaga or Benidorm.

GERALD GANNAWAY, *Bristol.*

Straight to the POINT

■ MUM was a fan of Andre Rieu, so we wanted his music at her cremation (Letters). However, we thought her favourite, Blaze Away, was not appropriate, so opted for 76 trombones.
SANDRA DRAPER, *Bexhill-on-Sea, E. Sussex.*

■ I ATTENDED the funeral of a prison officer where they played Jailhouse Rock.
F.M. LATTY, *East Bridgford, Notts.*

■ AT MY funeral, I intend to go out to the Four Tops singing I'm In A Different World.
JACKIE SHAW, *Theale, Berks.*

I CAN see how encouraging additional flights can add to climate deterioration. But I imagine the development of electric planes are a long way behind electric cars.

I'm often one of the unfortunates that, near the end of a 24-hour flight, spends another hour or two flying around in circles over London due to congestion at Heathrow. Has anyone factored in the impact of this on air pollution?
ERIC BREMNER, *Coventry.*

An eco lifestyle

I RECYCLE plastic and cardboard, save rainwater in butts for use on the garden, grass cuttings and vegetable peelings are made into compost and scrap wood from skips is turned into garden furniture and planters.

I even recycle the sunshine that falls onto my roof into electricity via solar panels.

When the eco-warriors each do as much as I do for the environment, I'll start to take them seriously.
TREVOR COLLINS, *Grimsby, Lincs.*

THE biggest problem with climate change is that you can't uninvent the wheel or put the genie back in the bottle.
ALAN LOCK, *Aylesford, Kent.*

■ IN MY case, no funeral music could be more apt than Elvis Presley's Way Down.
P. WILLIAMS, *Hayes, Middlesex.*

■ WHY doesn't Man City boss Pep Guardiola treat himself to a new cardigan. Or a defuzzer for the awful bobbly bits.
JANE BETTERIDGE, *Ashby-de-la-Zouch, Leics*

■ A PHOTO of our Eurovision entrant Michael Rice showed him waving an upside-down Union flag. Says it all!
CHRIS HORSMAN, *Little Stanton, Northants.*

Paying a high price

WHEN is the public going to wake up and realise we are being conned and overtaxed unnecessarily over climate change?

The revelation that solar farms received more in tax allowances than the value of the electricity they produce is ridiculous.

What about all those thousands of wind turbines, which have destroyed swathes of our countryside, but stand motionless because of not enough or too much wind?

Everyone is jumping on the bandwagon, from 'I am willing to believe anything' teenagers to 'I need some publicity' film stars and politicians.

The climate has fluctuated over the past 2,000 years. The Romans had vineyards in Northumberland and in the 18th century the Thames froze solid.

Calculations by climate change campaigners are flawed. They are based on guesswork regarding increases in population and long-range economic forecasts.

These same people predicted in the Seventies the world would run out of oil by 2000 and not so long ago told us all to buy diesel cars.

No one is saying we should do nothing, but sensible, well-thought-out plans are what is

266

A Virus in Society

From: Michael Cole
Sent: 13 May 2019 14:33
To: 'dtletters@telegraph.co.uk' <dtletters@telegraph.co.uk>
Subject: NO POINT IN SUCCESS

These eco zealots, now aided and abetted by some elements of Government, will ruin us all.

Our cars will be worthless as they are banned, and next our houses will have no resale value as they won't comply with the regulations.

We will not be able to eat what we like and our whole lives will be subject to control. We will become dependent on Government. Our work, companies and investments will be worth nothing unless it complies with eco-friendly regulations set by the fringe enthusiasts.

This may sound like something out of a Big Brother film but we are already on this path. The emperor's new clothes brigade are already influencing those gullible MPs on the back of unsound projections of doom about global warming.

We are so inward looking on this and it will lead to a different doom if we follow such unnecessary and extreme approaches. The developing world with their industrial leading economies will not slow down as they build wealth for their citizens and we, in the UK, will be left behind – who will support our eco saints then? It is foolish but who will be the 'heretic' in Government to challenge it and bring back some sense and balance to this issue?!

Regards,
Michael J Cole
Staffordshire, England

Michael J Cole

From: Michael Cole
Sent: 29 April 2019 12:05
To: 'letters@dailymail.co.uk' <letters@dailymail.co.uk>
Subject: PROTEST POLICIES

Why do our politicians and our Government allow the vocal minority protestors set policies, policies to damage our economy in knee jerk reactions? What about the silent majority, most of whom don't resort to twitter – we simply get ignored.

Richard Littlejohn's article in Friday's Daily Mail made the point exactly.

We let a protest group on climate change, represented by a 16 year old, have disproportionate influence over policies and the gullible politicians swallow the rhetoric in a bout of popularist enthusiasm. Are they so inept and uninformed that they just fall for it?

Now Corbyn, the popularist par excellence, will aim to put climate change at the heart of all investment moves. What complete nonsense. He and all his acolytes, including communist loving McDonnell, will largely do and say anything to get acceptance from the vulnerable and the youth. Why is there this preoccupation about youth? Lots of civilisations value the elderly, we used to but no longer. What about experience and judgement over misplaced bias and uncontrolled enthusiasm without knowledge?

All MPs need to give some serious consideration to these PC type moves and balance them with constructive evaluation.

We now have effectively abandoned our opportunity to benefit from gas fracking with the resignation of Natascha Engle, Shale Gas Commissioner. Our misguided Government has yielded to the minority objectors and "green zealots" by imposing impractical restrictions on the engineering to the detriment of our economy, the public at large and our Country.

With protestors to Trump's visit and bigots like Bercow and the Labour Party leaders we will be left as a Third World Country following these policies of economic collapse.

A Virus in Society

The sacking of Sir Roger Scruton was a disgrace and another example of lost freedom of speech and freedom of thought. Protests and challenges are now only allowed if it conforms to the PC brigade's ideas. Government is only too ready to go along with it but many of us don't, we wish to remain free!

What on earth's happening to Britain – something needs to change with our political leadership to reinforce common sense. Who will stand up for this?!

Regards,
Michael J Cole
Staffordshire, England

Michael J Cole

From: Michael Cole
Sent: 24 April 2019 09:23
To: 'letters@mailonsunday.co.uk' <letters@mailonsunday.co.uk>
Subject: ATTENTION - PETER HITCHENS

Dear Mr Hitchens

I totally concur with your article in the Mail on Sunday of 21 April about the left wing zealots and the climate change excuse for disrupting life for the hard working, largely middle England people. People who choose not to cause problems for the public and have better things to do.

Please find a couple of my latest correspondence to the media on this matter.

This nonsense and frankly exaggeration about the effects of manmade impact on our environment and the dire warnings reminds me of an old chap, a vagrant, who walked around Doncaster in the 1960's with a sandwich board (a two piece banner for those who don't remember them) stating 'The End of the World is Nigh' on one side and 'Put your Faith in Jesus' on the other. Some fifty odd years later 'nigh' hasn't happened!

Nor will it on the basis of floored projections – as a chemist background I have studied this. Why do we let these people have a disproportionate influence? The right to protest I fully support, my father and grandfathers fought two World Wars for the right to do so, but God forbid the weak popularist seeking politicians get influenced and are persuaded to implement any of these useless and made suggestions!

I am glad that you offer challenges to the rhetoric of these people and help provide a balance of opinion as does Richard Littlejohn in his own style. Please keep this up so that there is a chance that common sense will prevail.

Regards,
Michael J Cole
Staffordshire, England

A Virus in Society

From: Anglia Holdings [mailto:angliahldgs@btconnect.com]
Sent: 23 April 2019 13:23
To: 'letters@dailymail.co.uk' <letters@dailymail.co.uk>
Subject: EMOTIONAL BLACKMAIL

What a disappointing programme by David Attenborough, it was a series of claims by supporters of the Climate Change religion expanding doom with constructed images of emotional scenes. Scenes of a "dying bat with its baby" and cut into films of industrial plants.

The programme was entitled "The Facts" but there were very little if any, just statements from those in favour of the concept and the exaggerated consequences. Their fancy job titles did not persuade me of the robustness of their claims.

Now we even have the ridiculous proposals by some of the participants that we rid ourselves of all our vehicles and industry by 2025! The nonsense of this argument doesn't warrant comment.

We have seen this before with Al Gore's misleading film on this subject and the distorted and manipulated data from academic institutions who constructed the projected effects by adapting input data to give them the results they wanted. Hardly objective science! Bjorn Lomborg's book takes to task the misrepresented data by the global warmist community – it is worth reading.

And here we are again. No real facts in Attenborough's film just a repeat of the rhetoric served up to tug at the uninformed. The BBC, with its committed climate change bias is only too pleased to promote such distortions.

So, let's not get carried away with the emotional pitch by those committed zealots of the climate change religion and bring in a sense of balance. Let's also have some alternative assessments built on objective science and side step the irrational and impractical proposals which will damage our life and the quality of it.

Michael J Cole

Controlling waste and pollution is a reasonable focus but expanding this to eliminate industry, travel, mobility and effectively trade is not. Can you imagine this being followed by large parts of the world such as China, Asia in total and India when their economic prosperity is replacing poverty? We in the UK should not disadvantage ourselves by taking such unreasonable actions. If we do it will ruin our economy and create civil unrest. Not only climate change warriors can predict gloom!

There are challenges to this but they are rarely, if ever, heard, never on the BBC. The fact that the surface temperatures have fallen, Polar Bears are thriving and the ice caps are growing in areas are ignored. The earth within the cosmos is tiny and the main influences on climate and weather is the sun, sun spots and its cycle – not man made CO_2. After six billion or more years do we really think man can destroy our plant in effectively 100 years, as this is the industrial age claimed to be causing its downfall?

We live in the 21st century, we need a successful economy to provide for social issues and common sense needs to prevail. I'm happy to let my great, great grandchildren and beyond fend for themselves, the argument that we must 'provide' for them is totally out of context and arrogant.

It really is time a sense of proportion was applied, particularly to avoid over reaction by politicians in the knee jerk way they react to such matters (and see the opportunity to tax us more!).

Regards,
Michael J Cole
Staffordshire, England

EPILOGUE

This book outlines the dangers as seen by the writer of society falling into the clutches of doctrines designed to restrict freedoms and as indicated by many other publications. By definition George Orwell's book, '1984' predicted such dangers and it considered that some thirty odd years after the assigned year it is coming true.

The state and states are moving in that direction aided and abetted by technology and a section of the populous and media aligned in their will to make such changes and construct laws to support it and create 'operating principles' of behaviour to reinforce social structures.

To restrict and contain human thought and actions is the aim. Evidence of easy and effective containment can be demonstrated by the COVID-19 virus issue whereby whole countries are in lockdown as demanded by their controlling governments. These are reminiscent of earlier regimes, including in the 20th Century communist states and Nazism albeit that the basis was, arguably different. On the proposition of fear or 'saving lives' such moves are 'sold' to the public as necessary or essential.

These changes are real. They will consume society and individuals. The moves will affect all aspects of human life, make people dependent and subservient to the state (and wider collective states) so blunting ambition, independence and the very nature of humanity for an overreaching dogma of how people should live. It will and currently is creeping up on the masses in the developed world and will form a blueprint for the next generation.

This virus of containment of humanity will be, arguably, more virulent and damaging to human existence, hitting at the very core of what drives us, than an endemic organism as there will be no easy vaccination.

This may well be seen as a somewhat pessimistic or depressing analysis of the state of society's prospects and even unreasonably so. But changes and influences affecting society seen over the past generation or 25 years have been deeper and more fundamental than before in terms of political and intrusive effect on the way lives are governed and controlled. This is seen as more widespread than other impacts and more personal, even more than World Wars which had a more focussed and solvable recovery. It is, therefore, more concerning and it is reflected in all parts of individual lives.

Once this is fully established it will hopefully implode as sustainability of the status will be challenged by humankind's natural ambitions. Until then it is imperative that totalitarianism controls are seen and challenged now to stop them and tackle each element before they get greater hold and hopefully this book will help alert people to this virus in our society.

A FINAL THOUGHT

This is an extract from Michael Sangster's book and it summarises the positon of the principles of science and objectivity and how this should be the basis of understanding. It also shows that this has been distorted by vested interest groups and politics in the case of climate and changes to give wrong explanations and resultant wrong economic and social policies.

> **A QUOTE FROM DR MICHAEL SANGSTER'S BOOK**
> **"The Real Inconvenient Truth"**
>
> **"The alternative to uncertainty is authority, against which science has fought for centuries"**
> *Dr. Richard Fenyman, one of America's greatest scientists*
>
> **"The urge to save humanity is almost always only a face for the urge to rule it."**
> *H. L. Mencken, Journalist and Culture Critic*
>
> One of the most disturbing aspects of the entire global warming debate is the certitude of believers, driven almost entirely, if not exclusively by tribal affiliations.
>
> This shows that while CO_2 contributes to climate conditions it is not a primary driver, and that the climate models on which energy policies are based are invalid. It also addresses the unconscionable tactics employed by politicians and their media acolytes to perpetuate the myth that human caused CO_2 is responsible for global warming and climate change.

> A small group of ideologues have over the years hijacked an otherwise noble cause, environmental consciousness, and used it as a proxy to redistribute wealth and ultimately to replace Western style democracies with a world socialist structure under the auspices of the United Nations. To achieve their goals they use Alinsky-style tactics to denigrate, ridicule and intimidate skeptics and Orwellian methods, including rewriting history, to ensure conformity to their message.
>
> There are no angry mobs demanding consensus for the theory of Gravity $E=mc^2$ because they are testable, repeatable and confirmed through experiment and observation.
>
> The un-testable, un-provable relationship between CO_2 and a warming planet plays into the hands of scrupulous ideologues. Only through such an oblique association could they develop a pseudo-secular religion with millions of believers who view those who question its dogma as heretics.

This need for objective and balanced approaches to other wider social issues can also be made. Through this book it is hoped the misdirected approaches can be exposed and realigned for the sake of humanity.

Lightning Source UK Ltd.
Milton Keynes UK
UKHW010919080223
416610UK00014B/1264